Drought in Brazil

Proactive Management and Policy

Drought and Water Crises

Series Editor: Donald A. Wilhite

Published Titles:

Drought in Brazil: Proactive Management and Policy
Editors: Erwin De Nys, Nathan L. Engle, and Antônio Rocha Magalhães

Water Policy and Planning in a Variable and Changing Climate
Editors: Kathleen A. Miller, Alan F. Hamlet, Douglas S. Kenney, and Kelly T. Redmond

Drought, Risk Management, and Policy: Decision Making under Uncertainty
Editors: Linda Courtenay Botterill and Geoff Cockfield

Remote Sensing of Drought: Innovative Monitoring Approaches
Editors: Brian D. Wardlow, Martha C. Anderson, and James P. Verdin

Drought in Brazil

Proactive Management and Policy

Edited by Erwin De Nys
Nathan L. Engle • Antônio Rocha Magalhães

CRC Press
Taylor & Francis Group
Boca Raton London New York

CRC Press is an imprint of the
Taylor & Francis Group, an **informa** business

International Bank for Reconstruction and Development/The World Bank

CRC Press
Taylor & Francis Group
6000 Broken Sound Parkway NW, Suite 300
Boca Raton, FL 33487-2742

First issued in paperback 2019

ISBN-13: 978-1-4987-6566-4 (hbk)
ISBN-13: 978-0-367-87795-8 (pbk)

Library of Congress Cataloging-in-Publication Data

Names: Engle, Nathan, author. | De Nys, Erwin author. | Magalhäaes, Antonio Rocha, author.
Title: Drought in Brazil : proactive management and policy / Nathan Engle, Erwin De Nys, and Antonio Rocha Magalhaes.
Description: Boca Raton, FL : Taylor & Francis, 2016. | Series: Drought and water crises | Includes bibliographical references and index.
Identifiers: LCCN 2016005747 | ISBN 9781498765664 (hard copy)
Subjects: LCSH: Droughts--Brazil, Northeast. | Drought management--Brazil, Northeast. | Droughts--Government policy--Northeast. | Water supply--Brazil, Northeast.
Classification: LCC QC929.28.B6 E64 2016 | DDC 363.34/9295610981--dc23
LC record available at https://lccn.loc.gov/2016005747

Visit the Taylor & Francis Web site at
http://www.taylorandfrancis.com

and the CRC Press Web site at
http://www.crcpress.com

Contents

Foreword vii
Acknowledgments ix
Editors and Authors xi

1. **Life and Drought in Brazil** 1
   Antônio Rocha Magalhães

2. **Crisis, Opportunity, and Leadership** 19
   Eduardo Sávio P.R. Martins, Francisco José Coelho Teixeira, João Gilberto Lotufo Conejo, José Machado, and Antonio Divino Moura

3. **A Framework and Convening Power** 27
   Erwin De Nys, Nathan L. Engle, and Carmen Molejón Quintana

4. **The Technical and Institutional Case: The Northeast Drought Monitor as the Anchor and Facilitator of Collaboration** 37
   Eduardo Sávio P.R. Martins, Carmen Molejón Quintana, Maria Assunção F. Silva Dias, Robson Franklin Vieira Silva, Bruno Biazeto, Gisela Damm Forattini, and Julia Cadaval Martins

5. **Water Supply and Management Drought Preparedness Plans** 49
   Francisco de Assis de Souza Filho, Marcos Thadeu Abicalil, Paula Pedreira de Freitas de Oliveira, and Cybelle Frazão Costa Braga

6. **Municipal Agricultural Drought Preparedness and Response Plan** 67
   Barbara Farinelli, Pablo Valdivia, and Diego Arias

7. **Perspectives from the Outside: Contributions to the Drought Paradigm Shift in Brazil from Spain, Mexico, and the United States** 81
   Michael Hayes, Mario López Pérez, Joaquín Andreu, Mark Svoboda, Brian Fuchs, Felipe Arreguín Cortés, and Nathan L. Engle

8. **Planning for the Next Drought and Paving the Path for Climate Change Resilience** 91
   Nathan L. Engle, Erwin De Nys, Antônio Rocha Magalhães, and John Redwood III

9. **Voices of the People: Socioeconomic Implications of Drought in Northeast Brazil** 97
Dorte Verner

10. **Drought Impacts and Cost Analysis for Northeast Brazil** 119
Paulo Bastos

11. **Northeast Drought Monitor: The Process** 143
Eduardo Sávio P.R. Martins, Robson Franklin Vieira Silva, Bruno Biazeto, and Carmen Molejón Quintana

12. **Drought Preparedness Plans: Tools and Case Studies** 167
Francisco de Assis de Souza Filho, Paula Pedreira de Freitas de Oliveira, Marcos Thadeu Abicalil, Cybelle Frazão Costa Braga, Samíria Maria Oliveira da Silva, Sandra Helena Silva de Aquino, Daniel Antonio Camelo Cid, Luiz Martins de Araújo Júnior, and Ana Cláudia Fernandes Medeiros Braga

Bibliography 211

Index 221

Foreword

This book is part of an ongoing series published by CRC Press under the title "Drought and Water Crises: Science, Technology, and Management Issues." In 2005, I edited a book with the same title that addressed the intersection of drought and water-related issues. Since the publication of that book, concerns have continued to mount around the linkages between drought and water issues such as climate change, water scarcity, food security, and the sustainable use and management of natural resources. These are topics of frequent concern and debate in the scientific literature, the policy arena, and the media. The opportunity to expand discussion on these and other topics through a series of books aimed at drought and water management is timely and will assist scientists, natural resource managers, and policy makers in gaining a better understanding of these important issues.

Drought is a slow-onset natural hazard that is often referred to as a creeping phenomenon. The challenge of monitoring drought's onset, evolution, and identifying its termination or end is one that scientists, natural resource managers, and decision makers have been struggling with for decades. However, drought management must be aimed at reducing the risks of future drought events on economies, the environment, and the social fabric of regions.

Erwin De Nys, Nathan L. Engle, and Antônio Rocha Magalhães have edited a remarkable book for the next addition to this book series. *Drought in Brazil: Proactive Management and Policy* provides an important story of recent efforts in Brazil to move the country away from crisis management to proactive management of droughts in order to demonstrate how development practitioners and government officials might achieve similar paradigm shifts in other countries. Over my career, I have had the opportunity to work with Brazilian officials in pursuit of this goal. Strides have been made over the past several decades, but recent efforts have been more dramatic, borrowing experiences and technologies from the United States, Mexico, Spain, and other countries. This progress was, to a large degree, stimulated by a series of extreme drought years that affected Brazil, especially the Northeast region.

The editors have structured this book in a manner that is intended to appeal to various audiences. The recent advancements in drought management and policy in Brazil, while are in part very technical and scientific in nature, are first told in more of a story-telling format (Chapters 1 through 9) that will appeal to policy makers and development practitioners alike. The more technical details are elaborated upon in the second part of the book (Chapters 10 through 12). In the end, this book presents valuable lessons for all drought-prone regions, as they struggle to meet the real-world challenges of drought management today and in the future, given our changing climate.

Donald A. Wilhite
University of Nebraska-Lincoln
Lincoln, Nebraska

Acknowledgments

We express our most sincere appreciation to all the people involved with activities that inspired the publication of this book; most importantly, the countless individuals and institutions involved with planning for and managing droughts in Northeast Brazil.

Our deepest gratitude goes to a few prominent people in Brazil who seized the opportunity present during the critical drought that the Northeast region has been going through since 2010 and took bold actions to move toward a more proactive risk-based management and planning for droughts in the country. Francisco José Coelho Teixeira, first as the National Secretary of Water Infrastructure during 2012–2013 and then as the Minister of National Integration during 2013–2015, formulated a vision for a National Drought Policy that stimulated all those involved in the activities described in this book. José Machado, as the adviser to Mr. Teixeira during his tenure at the Ministry of National Integration, was influential in translating this vision into concrete institutional and technical building blocks, and channeling strategic support from Brazilian institutions and development partners. Eduardo Sávio P.R. Martins, the president of the Foundation for Meteorology and Water Resources of the state of Ceará, provided the methodological leadership that inspired all the technical teams, and was instrumental in forging the institutional coordination needed among the nine states of Northeast Brazil, and between the federal and state levels, to produce the Northeast Drought Monitor. We are grateful to Deborah L. Wetzel, country director for the World Bank in Brazil, during the execution of the technical collaboration program described in this book and for her encouragements to support Brazil with best international practices for drought management and planning.

The production of this book and the activities described herein were made possible thanks to the generous financial support of the World Bank's Water Partnership Program and the Spanish Fund for Latin America and the Caribbean. John Redwood III interlaced a common voice across Chapters 1 through 9, and supported the editors with translation services and multiple reviews of the manuscript. Julia Cadaval Martins' coordinating role cannot be overstated, as she was pivotal in keeping the editors on track throughout the book's production process, particularly in ensuring completeness and quality of the submissions from the various authors. Carolina Abreu dos Santos provided the logistical support and Vinícius Cruvinel Rêgo developed many of the maps and figures used in this book.

Editors and Authors

Marcos Thadeu Abicalil is a senior water and sanitation specialist at the World Bank in Brasília, Brazil. He has worked extensively on the Brazil integrated water and urban portfolios, with emphasis on the Northeast region of Brazil, leading or contributing to several investment and policy development lending operations to federal, state, and municipal governments.

Joaquín Andreu is a full professor and director of the Department of Hydraulic and Environmental Engineering at the Universidad Politécnica de Valencia (UPV), Spain. He has been the technical director of the Jucar River Basin Agency (CHJ). Joaquín specializes in integrated water resources planning and management (IWRPM), decision support systems for integrated water resources management (IWRM), drought risk assessment and management, and water conflict resolution.

Diego Arias is a senior agriculture economist at the World Bank in Brasília, Brazil. He works on agriculture development operations in the country and does analytical work on agriculture risk management globally. Diego has a PhD in agrifood and environmental economics from Universidad Politecnica de Valencia (UPV), a master's in international development from Harvard University, and an MBA from the Sacred Heart University in Luxembourg.

Felipe Arreguín Cortés is the general director of the Mexican Institute of Water Technology and a professor in the Division of Postgraduate Studies, Faculty of Engineering, National Autonomous University of Mexico. He is also the vice-president of the Association of Civil Engineers of Mexico and a member of the Pan-American Academy of Engineering. His PhD research is on hydraulics.

Francisco de Assis de Souza Filho is a professor at the Federal University of Ceará and coordinates the Water Resources and Environmental Hydraulics Program of the Department of Environment. Previously, he was the president of the Brazilian Water Resources Association and the president of the Ceará Foundation for Meteorology and Water Resources (FUNCEME).

Paulo Bastos is a senior economist in the Development Research Group at the World Bank in Washington, DC. His research interests are broad, including development economics, international trade, labor economics, and climate change. He holds a PhD in economics from the University of Nottingham.

Bruno Biazeto is a founding partner of ARESTech, a consulting firm in the environmental area. He is involved with several projects and collaborations with MASTER Laboratory/USP and generally works on geosciences, mesoscale meteorology, and atmospheric modeling and automation of forecasting systems. Bruno has bachelor's and master's degrees in meteorology from IAG/USP.

Julia Cadaval Martins is a doctoral candidate at Georgetown University in Washington, DC, focusing on comparative water governance and development. She works as a governance and institutions specialist at the World Bank, and previously acted as a consultant on industry regulation and public–private partnerships in Brazil. She holds an LLM (master's) degree from Harvard Law School, a law degree from Rio de Janeiro State University, and a master's degree from PUC-Rio.

Daniel Antonio Camelo Cid is a master's student in civil engineering with emphasis on water resources at the Federal University of Ceará, and holds a bachelor's degree in civil engineering from the Federal University of Ceará. He is a researcher at climate risk management for water sustainability research group—UFC-DEHA.

Erwin De Nys is a program leader for sustainable development at the World Bank, based in Bogotá, Colombia. During 2011–2015, he coordinated the World Bank's portfolio in Brazil related to analytical work and investment projects in water resources management, irrigation management, and adaptation to climate change. Erwin was a co-leader of the "Drought Preparedness and Climate Resilience" technical collaboration program described in this book. He has a PhD in irrigation management from the University of Leuven—KU Leuven (Belgium), a master's in water resources management from the International Centre for Higher Education in Agricultural Sciences (Montpellier, France), and a master's in tropical agriculture from the University of Leuven—KU Leuven (Belgium).

Nathan L. Engle is a climate change specialist at the World Bank in Washington, DC. His work focuses on climate resilience and adaptation, drought policy and management, and water resources. Nathan was a co-leader of the "Drought Preparedness and Climate Resilience" technical collaboration program described in this book. He holds a PhD from the University of Michigan's School of Natural Resources & Environment, master's degrees from the University of Michigan in public policy (MPP) and natural resources and environment (MSc), and a BSc in earth sciences from the Pennsylvania State University.

Barbara Farinelli is an agricultural economist at the World Bank in Brasília, Brazil. She co-manages investment operations and analytical studies in agriculture and climate change topics. Barbara holds a master's degree in agricultural economics and natural resources from the University of California, Davis.

Ana Cláudia Fernandes Medeiros Braga is an assistant professor at the Federal University of Paraíba, Brazil, where she works on several projects related to climate and water resources, with emphasis on Northeast Brazil. She holds a master's in civil engineering (UFPB) and a PhD in natural resources (UFCG).

Gisela Damm Forattini is a water resources specialist and director at the National Water Agency (ANA) of Brazil. She was formerly the licensing director at the Brazilian Institute of the Environment and Renewable Natural Resources (IBAMA) and served as the chair of Global Water Partnership Brazil and Global Water Partnership South America. Gisela has a master's in water resources and sanitation.

Brian Fuchs is a climatologist and a faculty member for the National Drought Mitigation Center (NDMC) at the University of Nebraska in Lincoln's School of Natural Resources. He works on applied research projects for the center as well as co-authoring the "United States Drought Monitor" and the "North American Drought Monitor."

Cybelle Frazão Costa Braga is an assistant professor of education, science and technology at the Federal Institute of Paraíba (IFPB), Brazil, since 2013. She researches several projects in water resources management, including World Bank-financed projects on water and climate change. Cybelle is a civil engineer, and holds a bachelor's in law, a master's in civil engineering (UFPB), and a PhD in natural resources (UFCG).

Paula Pedreira de Freitas de Oliveira is a senior water resources specialist working in Brazil. She joined the World Bank in 1997 and, since then has been working on the water resources and irrigation portfolios, leading or contributing to studies, technical assistance, investments, and policy development lending operations to federal and state governments. Paula works closely with Brazilian counterparts, participating actively in the sector dialogue in the country.

Michael Hayes is the director for the National Drought Mitigation Center (NDMC) located within the School of Natural Resources at the University of Nebraska–Lincoln. His main interests deal with drought monitoring, planning, and mitigation strategies. Michael received a bachelor's degree in meteorology from the University of Wisconsin–Madison, and his master's and doctoral degrees in atmospheric sciences from the University of Missouri–Columbia.

Sandra Helena Silva de Aquino is a doctoral student in sociology at the Federal University of Ceará (UFC), has a master's degree in development and environment-PRODEMA from UFC, and graduated in social sciences from the State University of Ceará. She is a collaborator in the team of water management and water sustainability (UFC-DEHA).

Mario López Pérez is a hydrology coordinator at the Mexican Water Technology Institute. He has worked over the past 34 years for the Mexican Federal Administration in the water sector, including most recently as the engineering and binational water issues manager at the National Water Commission of Mexico. He holds an agronomy engineering degree from the Universidad Autónoma Metropolitana; an MSc degree in pedology and soil survey from the University of Reading, the United Kingdom; and a diploma on public administration with honors from Universidad La Salle of Mexico.

João Gilberto Lotufo Conejo is the director of regulation at the National Water Agency (ANA) in Brazil and an assistant professor at the University of São Paulo. Previously, he was in charge of the water resources planning sector at ANA. He has been the head of the Water Resources Management Agency of the State of São Paulo; the deputy secretary of the water resources, sanitation and works secretary of the State of São Paulo; technical director of the Sanitation Municipal Company of Campinas; and an engineer of the Water Resources Management Agency of the State of São Paulo (DAEE/SP). He is a civil engineer and holds an MSc in water resources from the University of São Paulo and has received a diploma of membership of the Imperial College in Hydraulic Engineering at the University of London.

José Machado is an economist and special advisor to the President of the Company for the Development of the São Francisco and Parnaíba Valleys (CODEVASF). Formerly, he served as the president of Brazil's National Water Agency (ANA), the executive secretary of the Ministry of the Environment, and as a special advisor to the Minister of the National Integration.

Antônio Rocha Magalhães is an advisor at the Center for Strategic Studies and Management (CGEE) in Brasília, Brazil, dealing mostly with issues linked to droughts in the Northeast region, science and technology, water, and adaptation to climate change. Formerly, he was the secretary of planning of the State of Ceará and the vice minister of planning of Brazil, the president of the Committee of Science and Technology of the United Nations Convention to Combat Desertification, and a country officer at the World Bank in Brazil. He holds a PhD in economics from the University of São Paulo.

Eduardo Sávio P.R. Martins is a professor at the Federal University of Ceará (UFC) and the president of the Ceará Foundation for Meteorology and Water Resources (FUNCEME). His research focuses on hydrological extremes, use of weather/climate information in water resources, hydrologic statistics, and climate change impact assessments in water and agriculture sectors. He has a PhD in civil and environmental engineering from Cornell University, an MSc in water resources and environmental sanitation from the Federal University of Rio Grande do Sul, and a civil engineering degree from the Federal University of Ceará.

Luiz Martins de Araujo Junior is a PhD student in civil engineering with emphasis on water resources at the Federal University of Ceará. Previously, he worked in the Ceará Foundation for Meteorology and Water Resources (FUNCEME) at the Department of Meteorology (DEMET). He holds a degree in physics and a master's degree in physical sciences from the state University of Ceará.

Carmen Molejón Quintana is a water resources specialist at the World Bank, based in Brasília, Brazil. She is involved in different projects with the Brazilian federal government and in the Northeast region, and her experience is in water resources, rural water supply, geographic information systems, and international development areas. She has a degree in civil engineering from the University of Coruña, Spain.

Antonio Divino Moura is the director of the Meteorological Institute of Brazil (INMET) and the first vice-president of the World Meteorological Organization (WMO). He was a researcher and professor of meteorology at the Brazilian Space Research Institute (INPE). He has a PhD in meteorology from MIT and pursued postdoctoral studies at the NASA-Goddard Space Flight Center on droughts in Northeast Brazil and the role of oceans. He is an honorary fellow of the Royal Meteorological Society and member of the World Academy of Sciences, Trieste, Italy.

Samíria Maria Oliveira da Silva is a PhD candidate in civil engineering at the Federal University of Ceará with emphasis on management and economics of water resources and water allocation. She is a researcher at the Climate Risk Management and Water Sustainability research group.

John Redwood III is an independent international consultant on environment, climate change, and sustainable development. Formerly, he was a World Bank staff member and sector director, and has worked in various parts of Brazil for the US Peace Corps, two federal universities, the Organization of American States, and the Inter-American Development Bank. He holds a BA from Harvard University, and an MCP and PhD from the University of California, Berkeley.

Maria Assunção F. Silva Dias is a professor at the University of São Paulo, where her research focuses on rainfall extremes, thunderstorms, Amazon biosphere–atmosphere interactions, local circulations in urban areas, and aerosol–rainfall interactions. Formerly, she was the director of the Center for Weather Forecasting and Climate Studies in Brazil, and also deputy director of the Institute of Astronomy, Geophysics, and Atmospheric Sciences of the University of São Paulo.

Robson Franklin Vieira Silva is an IT specialist with the Ceará Foundation for Meteorology and Water Resources (FUNCEME), where he works on hydrologic systems, databases, radar meteorology, remote sensing, and weather/climate forecasting and monitoring applications. He holds a master's in applied science physics from the State University of Ceará (UECE).

Mark Svoboda is a climatologist and monitoring program area leader for the National Drought Mitigation Center (NDMC). He cofounded the development of the US Drought Monitor (USDM) in 1999 and serves as one of the principal authors of both the weekly "USDM" and monthly "North American Drought Monitor." Mark earned both his bachelor's degree in geography and his master's degree in geography with a specialization in remote sensing, climatology, and GIS from the University of Nebraska–Lincoln.

Francisco José Coelho Teixeira is a water resources secretary for Ceará State. Formerly, he was the minister of Brazil's Ministry of National Integration (MNI), the infrastructure secretary for MNI, and the director of the several water institutions of Ceará State (SOHIDRA and COGERH). He is a civil engineer and holds an MSc in water resources.

Pablo Valdivia is an agriculture insurance consultant at the World Bank, working mainly in Latin American and the Caribbean countries. His experience includes work with the public insurance sector in Nicaragua and as an agricultural business entrepreneur. He holds a BSc in agronomy from the EARTH University and is an MBA candidate at the University of Liverpool, the United Kingdom.

Dorte Verner is a lead economist in the agriculture sector at the World Bank in Washington, DC. Dorte has published books and papers in the areas of climate change, poverty reduction, and socioeconomic development. She holds a PhD in macroeconomics and econometrics from the European University Institute, Italy; and a postgraduate degree in economics from the University of Aarhus, Denmark.

1

Life and Drought in Brazil

Antônio Rocha Magalhães

CONTENTS

1.1 Northeast Brazil and the Semiarid Region 2
1.2 The Drought Problem in the Northeast 4
1.3 The Case of São Paulo 6
1.4 The Amazon Case 7
1.5 Social, Economic, Environmental, Political, and Cultural Impacts of the Drought in the Northeast Region 7
1.6 The Government's Response: A History of Reactive Policies 10
1.7 Emergency Responses to the Drought 14
1.8 The 2010–2015 Drought 16
1.9 The Droughts and Climate Change: An Emerging Threat 17
1.10 The Need for Proactive Responses 17

Brazil is a very large and diverse country, and this is reflected in its climate. At the same time as there can be frost in the South and Southeast of the country, temperatures can surpass 30°C in vast regions of the North and Northeast. Droughts can occur in all regions, from north to south and east to west, affecting agricultural production and water supply. However, it is in the Northeast where they happen with greatest frequency and intensity and have the most severe impacts. For this reason, this book focuses primarily on that part of the country.

This book defines drought as a sustained and regionally extensive occurrence of below average natural water availability. Drought can be seen as a deviation from the long-term conditions of variables such as precipitation, soil moisture, groundwater, and stream flow. Drought is usually caused by below average natural water availability due to climate variability, which results in low precipitation and/or high evaporation rates. However, it is important to distinguish drought from aridity and water scarcity. Aridity is a permanent feature of a dry climate, whereas drought is a deviation from the long-term climate. Drought is a natural phenomenon, but water scarcity occurs when the shortage of water is caused by humankind using more water than naturally available. Desertification is a more or less permanent degradation of land in semiarid and dry subhumid areas. Drought and water scarcity can contribute to desertification, but the main reasons are overgrazing, increased fire frequency, deforestation, and/or overabstraction of groundwater. All of these elements are present in the Northeast of Brazil, which is now also increasingly subject to the exacerbating impacts of climate change, as will be further discussed below.

1.1 Northeast Brazil and the Semiarid Region

The Northeast region is very large, covering an area of 1,561,177 km^2. It consists of nine states: Maranhão, Piauí, Ceará, Rio Grande do Norte, Paraíba, Pernambuco, Alagoas, Sergipe, and Bahia (from north to south). When one speaks of the *semiarid* region, the northern part of the state of Minas Gerais is also included (as shown in Figure 1.1).

The Northeast region, however, is not homogeneous. It has an extensive coastal area, where much of the population is concentrated, including a number of state capital cities (most notably Fortaleza, Recife, and Salvador). It rains significantly; therefore, it has plentiful vegetation, also known as the *forest zone* or *zona da mata*. In this humid coastal area, sugarcane, which for centuries was the principal economic activity of colonial Brazil, is continued to be planted. Cocoa also comes from this zone. In extreme drought situations, this area may be affected, but, in general, it has been largely spared, at least until the present time, from the effects of the drought.

Further inland along the eastern part of the Northeast region is the *agreste*, which is a large transition area between the *zona da mata* and the semiarid region. This area produces food crops and small-scale livestock and frequently suffers from droughts. Still further inland is the semiarid region, or *sertão*, an even more extensive area that normally suffers from water scarcity and is the part of the Northeast region that is strongly affected by periodic droughts. There is also significant food production, especially beans, corn, and cassava, in this region, as well as traditional livestock, including cattle, sheep, and goats. When one speaks of drought in the Northeast region, one is normally referring to the semiarid region or *sertão*. This involves an area of 982,563 km^2 where some 22.6 million people live and covers parts of eight northeastern states (from Piauí to Bahia) plus northern Minas Gerais (CGEE 2015). The semiarid region has been denominated the *drought polygon* (*Polígono das Secas*), which also includes parts of the *agreste* (as illustrated in Figure 1.2).

To the west and northwest of the semiarid area lies another large region in the states of Bahia, Piauí, and Maranhão that is an extension of the *cerrados* or savannas of central western Brazil. This is a recently occupied area and is now making a significant contribution to national grain production—especially soybean and corn. It is also less subject to droughts.

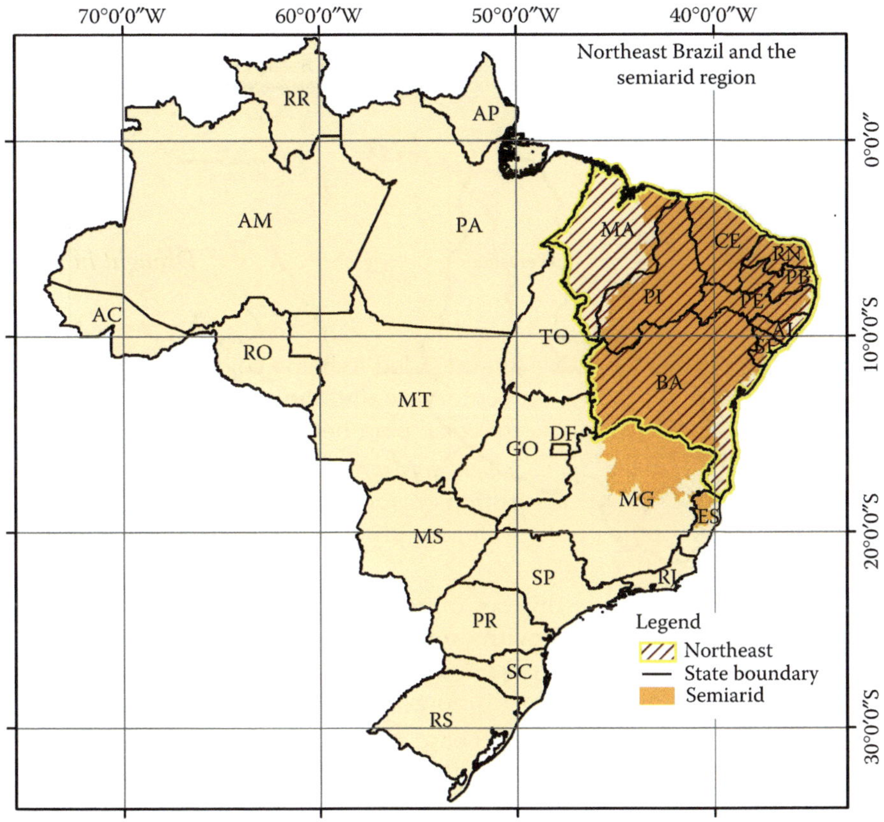

FIGURE 1.1
Map of Brazil and states of the Northeast region. (Data from Fundação Cearense de Meteorologia e Recursos Hídricos – FUNCEME, Fortaleza, Ceará, Brazil, 2015. Based on IBGE, 2007; MMA, 2007.)

North and west of the semiarid area, there is another transition zone called *pre-Amazônia* in the state of Maranhão, where, as in the *zona da mata*, tropical vegetation predominates. Here too, similar to the coastal zone, there is significant precipitation and, even though rainfall declines when droughts occur in other parts of the Northeast region, the effects on local output and society are less severe. The various subregions of the Northeast can be seen in Figure 1.2.

Finally, it is important to refer to the *caatinga* biome. This contains the lowland savanna-type vegetation that is typical of the semiarid zone, even though it is not present everywhere in this region. Certain other microclimates also exist in the semiarid region, such as the humid peaks (or *serras úmidas*), where vegetation types are different from the *caatinga*. For this reason, the map of the *caatinga*, which covers 826,411 km^2, does not coincide exactly with that of the semiarid zone or the Northeast as a whole, as can be seen in Figure 1.2.

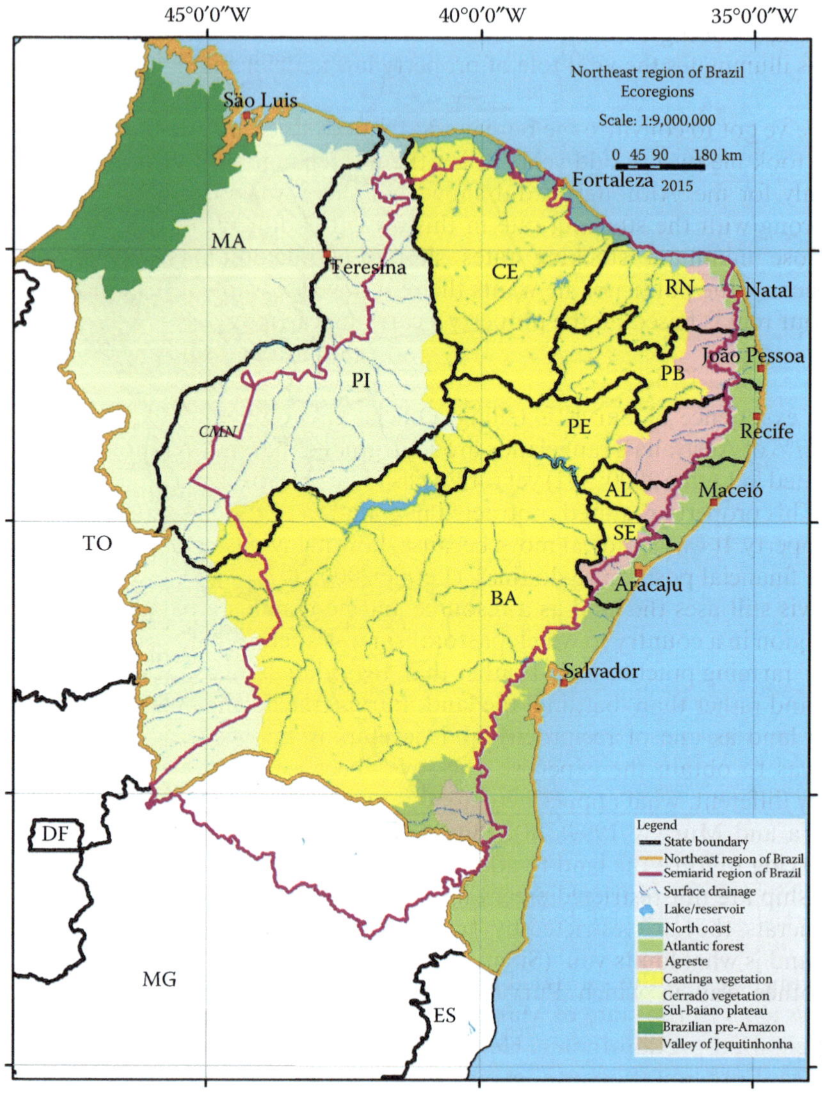

FIGURE 1.2
Major eco-regions of Northeast Brazil. (Data Fundação Cearense de Meteorologia e Recursos Hídricos - FUNCEME, 2015. Based on: USGS, 2005; CPRM, 2006; IBGE, 2006; IBGE, 1985; IBGE, 2007; BNB/FUNCEME, 2005.)

1.2 The Drought Problem in the Northeast

Droughts have historically influenced all aspects of life in the Northeast region. Even though they occur with greatest intensity in the semiarid part of the region, followed by the *agreste*, all of the Northeast region can be affected. In fact, the rivers that flow to the coast almost always have their origins in the *sertão*. For this reason, in this book we use the expression *drought in the Northeast* and *drought in the semiarid region* synonymously even though the two areas are not exactly the same, as explained above.

Droughts have always existed in the Northeast region. Prior to the settlement in the interior of the region that started in the middle of the sixteenth century, there were no significant problems because the predominant *caatinga* ecosystem was adapted to the climate and its periodic variations. The older and sparsely settled indigenous peoples who lived in the region were affected only in cases of extreme droughts, causing them to migrate toward the coast, according to those who chronicled the early years of Portuguese colonization. However, once new settlers penetrated the *sertão* and began to alter the landscape by establishing farms and cutting the native vegetation to raise livestock and produce crops, vulnerability to droughts gradually increased.

Over the years, various severe droughts have been recorded. The most devastating among them, in 1877–1879, killed half the human population and nearly all the cattle in the region. Prior to this, other strong droughts were registered, also causing significant impacts that tended to increase as human settlement expanded to the interior regions. After 1877, other major droughts occurred in 1900, 1915, 1919, 1932, 1942, 1951–1953, 1958, 1970, 1979–1983, 1987, 1990, 1992–1993, 1997–1998, 2002–2003, and 2010–2015 (CGEE 2015).

The causes of the periodic droughts are climate-related. The Northeast region's climate is strongly influenced by the El Niño phenomenon and by the surface temperatures of the Atlantic Ocean. The region also suffers from the influence of cold fronts that come from the South and winds that bring humidity from the Atlantic. In general, El Niño years are characterized by below-average rainfall in the semiarid region, which otherwise averages about 800 mm annually. Although comparatively high, this precipitation is concentrated during a few months of the year only. In addition, annual evapotranspiration levels exceed 2000 mm, which, associated with the shallow soils over a crystalline base in much of the semiarid region, results in its rivers being intermittent. During extreme droughts, rainfall decreases by more than 50%.

As the semiarid region is a frontier area in terms of climate, any reduction in average rainfall levels can provoke large impacts. Before human occupation, the untouched ecosystem was resilient and adapted to climate variations. However, the new situation created by increasing human settlement and land use has made the region progressively more vulnerable. A drought means lack of water for agriculture, for human consumption, and for both domestic and wild animals. The impacts are also economic (with the loss of agricultural production and livestock), social (with increased unemployment, hunger, and in extreme cases, fatalities among people who, in desperation, often seek to migrate temporarily or permanently in hopes of finding better places and more secure livelihoods), and environmental (with the death of wild animals, exhaustion of water sources, and growing land degradation and desertification, especially where there has previously been human interference in the form of deforestation and the clearing of vegetation for various purposes).

The problem is that, in the history of the Northeast and the semiarid regions, in particular, the drought is often forgotten once the phenomenon passes. People return to their homes in the *sertão* and use the land for agriculture, pasture, and extraction of firewood. Agricultural output rises and the drought problem is overlooked, until the next one begins and the process starts all over again. What happens here is now referred to as the *hydroillogical cycle*, an expression created by Professor Donald Wilhite of the University of Nebraska (Wilhite et al. 2005). The drought and semiaridness, in reality, are permanent components of the scenario in the interior of the Northeast region. Human activity needs to adapt to the conditions of the semiarid region, not the reverse.

1.3 The Case of São Paulo

Even though droughts are primarily associated with the Northeast region, they occur elsewhere in Brazil as well. However, their characteristics and effects are different due to the greater economic and social vulnerability of the Northeast region. But with rapid increase in population of the largest cities in the Southeast region over the past half century, especially in the São Paulo, Rio de Janeiro, and Belo Horizonte metropolitan areas, droughts also have created a significant impact on the residents. In fact, the availability of secure water sources for the metropolitan region of São Paulo with its nearly 20 million inhabitants has challenged decision makers in the state for some time. At present, the supply of water to this area depends on rainfall that fills local river basins and reservoirs, together with the importation of water from neighboring basins.

Some time ago, the Cantareira system (named after a local highland area within the metropolitan area) was built to import water from a river basin that supplies other parts of São Paulo. The water in the Cantareira system is stored in reservoirs that supply the metropolitan area. Before the 2014 São Paulo drought, nearly 70% of the water demand in metropolitan São Paulo was met by this system. But due to this drought, local reservoirs reached a very low level, causing one of the largest water supply crises experienced to date in São Paulo state. Part of the population normally attended by the Cantareira system had to be supplied by other systems, which lack space for expansion. The population of metropolitan São Paulo had to learn to live with less water, some economic activities needed to relocate, and the water supply crisis became a significant political problem, both locally and nationally. It may take several years for water in the reservoirs to return to normal levels, and this will depend on the intensity of the next few rainy seasons. Other aspects of the São Paulo case are discussed in Chapter 2.

1.4 The Amazon Case

The year 2005 was also historic, in that it recorded an extreme drought in Brazil's Amazon region. Even though the entire region suffered, the greatest impacts occurred in the southern part of the Amazon, below the Equator. A large area was affected by the drought, not only in Brazil but also in parts of Bolivia and Peru. Except for airplanes, the only means of transportation in many parts of the Amazon is river navigation. When a drought occurs, navigation becomes inviable in many localities, houses become distant from the receding riverbanks, towns are less easily supplied, and wells dry up. In the state of Amazonas alone, 137,000 families were affected by the 2005 drought and more than 100,000 tons of fish production was lost. Impacts were likewise very severe in the state of Acre. This drew the attention not only of the Brazilian government but also of international organizations such as the World Bank, which promoted a seminar to discuss the matter (World Bank 2005).

Other strong droughts were registered in the Amazon in the past, as in 1926 and 1963, and in many ways were even more severe (World Bank 2005). There was another strong drought in 2010, which particularly affected the Amazon, Solimões, and Negro River valleys. The water level fell to such an extent that various places where boats were anchored became dry land. More than 60,000 families were affected, food prices increased, and navigation was profoundly altered (Balzam 2010). The recent and apparently more frequent droughts in the region have been associated with the rising impacts of global climate change and the associated risk of what scientists have referred to as an *Amazon dieback* (Vergara and Scholtz 2011). Increased climate variability may also lead to greater frequency and severity of droughts in other parts of Brazil, especially the semiarid Northeast region, where its socioeconomic and environmental impacts have traditionally been the greatest, as will be discussed further below. In fact, during 2005, about one-third of Brazil was under drought conditions, including the Amazon, the Northeast, and the South regions (Taddei and Gamboggi 2010).

1.5 Social, Economic, Environmental, Political, and Cultural Impacts of the Drought in the Northeast Region

Throughout history, droughts have caused strong social, economic, and environmental impacts in the Northeast region. Their political and cultural consequences should also be highlighted. Once it is apparent that a year will be a dry one, that is when the rainfall is not sufficient to ensure a normal harvest and to store sufficient water in the reservoirs, farmers decide not to plant. Millions of smallholders and rural workers can find themselves unemployed, without an opportunity to work. A social calamity sets in, but before emigrating, residents of the semiarid region try everything possible to stay where they are. In less severe droughts, they can often temporarily find alternative sources of livelihood in the areas where they live. However, when droughts are more severe and/or extend for longer periods of time, they need to move to the larger towns and cities in the region, or to other parts of the country, including the Amazon, São Paulo, Rio de Janeiro, and Brasília. This has occurred numerous times in the past, for example, in 1877, 1915, 1932, 1958, 1970, and 1983.

In the nineteenth century and earlier, there were neither paved highways nor support systems. Drought refugees, known as *flagelados* or *retirantes* in Brazil, had to migrate on foot, sometimes for several months, and frequently died along the way. The American naturalist Herbert Smith, who visited Ceará, the northeastern state that has the largest share of its territory in the semiarid region, reported that some 500,000 people died in that state alone because of the 1877–1879 drought, as well as another 300,000 in other parts of the Northeast region (Smith 2012). A few years earlier, in 1872, Brazil had undertaken its first demographic census. At that time, Ceará had about 700,000 inhabitants, mostly living in the interior. Thus, extreme drought caused the death of more than half of the state's population. This disaster was not only reported in Brazil, but internationally (Cooper-Johnston 2000; Davis 2000; Smith 2012). Strong social impacts continued to occur in subsequent severe droughts, for example, in 1900, 1915, 1919, 1932, and 1958 among others, including unemployment, hunger, thirst, malnutrition, and death.

In 1915, and later in 1932, government authorities decided to create refugee camps to house drought refugees (Neves 1995). Thousands of drought displaced people were confined to these shelters thus avoiding, or at least reducing, pressure on the larger towns and cities, which were not prepared to receive them. In reality, this was also a calamity. Even as recently as 1958, a variation of this strategy was applied. In that year, the author of this chapter visited a facility known as the Getúlio Vargas Camp (or *Hospedaria* in Portuguese) in Fortaleza, where *retirantes* were confined waiting for an opportunity to migrate to Amazônia and hoping for the drought to end. It was a scene of pain and desperation.

As concerns economic impacts, nearly the entire livestock population in the drought-affected areas was decimated in 1877–1879, while food grain production, especially beans and corn, was almost totally lost. The economy of the semiarid region was practically destroyed during that drought. No survey of the great drought's environmental impacts was made at the time, but it is known that many trees failed to resist the lack of humidity and, as still occurs today, winds swept away the unprotected soil cover in the form of dust. Water sources dried up, depriving both people and animals access to this vital resource. Environmental impacts rose as the result of subsequent droughts, with larger areas having been deforested, more soils impoverished, and further riverine vegetation destroyed by the expansion of human occupation in the interim. Under these circumstances, droughts contribute to augment the intensity of land degradation and desertification. At present, this increasing degradation and desertification represent a growing threat to livelihoods in the semiarid region (CGEE 2015).

Droughts in the Northeast region are also deeply ingrained in regional culture. In general, anyone 10 years of age or older has already experienced one or more droughts. Consequently, drought is part of the region's history, music, literature, beliefs, and religiosity. Several of the most renowned artists and authors from the Northeast region have focused on the droughts and their impacts. Writers like Jorge Amado, Rachel de Queiroz, Rodolpho Theóphilo, Domingos Olímpio, and Graciliano Ramos dedicated important works to these themes. Perhaps the best known of which is a novel entitled *Vidas Secas* (*Dry Lives*) by the latter author. Artists like Luiz Gonzaga, the symbol of northeastern music, made the drought the predominant element in their compositions. Numerous popular poets have immortalized the droughts in their poems, the most famous among them, Patativa do Assaré. The Northeast region has also produced an ample popular literature called *literatura de cordel* because the booklets were hung from cords and sold in the markets in the towns and cities in the *sertão*, in which the drought is an ever-present theme. In short, it is a part of everyone's life.

And what can be said about the political aspect? Without a doubt, the political impact of the droughts depends on the type of predominant social relations in the semiarid region. Historically, everyone was affected, but in a light drought landowners were harmed less than landless rural workers. Still today, the predominant activity of the larger landholders, cattle raising, is generally impacted less than food production, which is carried out primarily by small farmers. Thus, larger producers are often less severely affected than smaller ones. In the past, many rural laborers lived on the larger land holdings and worked for the owners. At the start of a drought they often received some protection from the large landholders. But when drought conditions became more severe, they were left to their own devices.

This social system, in which rural workers depended heavily on the landowners, reflected the political situation. The large landholders, in practice, were also the local political bosses and controlled the workers' votes as part of the statewide political system. A strong drought, therefore, could disrupt this arrangement. Following a drought, the prevailing system could not always be reestablished. This explains, for example, why it was possible for a change in the political structure in the state of Ceará in 1987. As the power of the traditionally dominant large rural landowners weakened, the urban industrial class increased. Thus, drought impacts affected the political system and were also influenced by it.

The most significant issue surrounding the droughts is the question of water, which is the source of all the other impacts. Drought, in fact, means lack of water for human and animal consumption, for agriculture, and for other socioeconomic activities in the semiarid region. Its social, economic, and environmental impacts can be gradual, as rainfall diminishes and water sources become more scarce, first for human consumption, then for other purposes. People cannot survive long without water. For this reason, drought impacts can be so serious in so many realms.

1.6 The Government's Response: A History of Reactive Policies

From the time when a new drought begins in the Northeast region until the moment that it becomes a topic for government decision making, there is a certain delay. Sometimes, in less severe droughts, the impacts are not sufficient to provoke a public sector reaction. However, in strong droughts, the clamor by the affected population becomes a political problem and ends up arriving at the highest national decision levels. Prior to 1899, Brazil was a centralized empire, with decisions made in Rio de Janeiro, 2000 km from the Northeast region. When the government reacted to droughts like the one in 1845, it did so in a purely short-term, welfare-oriented (*assistencialista* in Portuguese), and insufficient way. In 1859, the imperial government organized a scientific commission to study the Northeast region and the drought problem (Braga 2004). This commission recommended the importation of camels from Africa to facilitate transportation in the semiarid region. The camels did not adapt and the experiment failed, but this consumed considerable time.

During the 1877–1879 drought, the local outcries and its duration were such that substantial concern was raised at the center of power. The actions that followed, however, were only *assistencialistas* in nature and once again were insufficient. During this period, political interference in the programs to assist drought victims was detected and created another type of problem: while government actions were taken with the intention of helping those most harmed by the drought, their administration was controlled by local political elites, who frequently took advantage of these measures for their own benefit. This problem of elite capture continued during succeeding droughts and remains a concern when decision systems are not transparent.

After the 1877–1879 great drought, when both the regional economy and a large share of its population had been decimated, the imperial government established a commission of engineers to develop a more permanent response to the drought. This commission recommended the construction of dams and reservoirs, known as *açudes* in the region, to store water and guarantee its supply during drought years. It also recommended the transposition of water from the São Francisco River, which is perennial, to rivers in Ceará, which are temporary ones (i.e., dry up for months even during normal rainfall years). The São Francisco River has its origins in the highlands of the state of Minas Gerais and flows northward through Bahia before turning eastward and forming the north–south boundary between the states of Bahia and Pernambuco and, further to the east, Sergipe and Alagoas, then flowing into the Atlantic Ocean. It thus runs through the heart of the semiarid region and has been the most important internal waterway in the Northeast region since colonial times. Construction on the first large *açude* began in 1886 when Brazil was still a constitutional empire, but was not completed until 1906, at which point the country had become a republic.

The strategy to build *açudes* continued to the present day, being reinforced every time a new drought occurred. In short, the same solutions were triggered by new disasters and never in a proactive forward-looking way. The civil works to permit the transposition of water from the São Francisco basin are, at the time this book was written, still in the construction phase, nearly a century after their initial recommendation by the imperial engineers. According to the Ministry of National Integration (MNI), this project for the northern part of the Northeast region will involve the construction of 700 km of canals and should benefit 12 million people in 390 municipalities (MNI 2015). When it will be completed, however, is not presently clear, although the first part of the system is planned to be operational in 2016.

An important new step in official responses occurred in 1909 when the federal government created a permanent institution, the Inspectorate of (Public) Works against the Drought. IOCS, as it was known at the time, was modeled on the United States Bureau of Reclamation. Later, in 1919, IOCS was transformed into IFOCS, Federal Inspectorate of Works against Droughts, and in 1945, IFOCS was transformed into the present-day DNOCS, the National Department for Works Against Droughts. DNOCS has continued to implement the policy of dams and reservoirs (or *açudagem*) and also to undertake studies to improve knowledge about the semiarid region and the *caatinga*.

The *hydraulic phase* of government response to the droughts in the Northeast region continued to dominate policymaking until 1945. However, starting in that year, new institutions were created, reflecting a broader understanding that, in addition to building physical structures, it was necessary to promote socioeconomic development and to generate employment in the region. Still in the 1940s, the Superintendency for the São Francisco Valley was established and modeled after the Tennessee Valley Authority (TVA) in the United States. It is known today as CODEVASF, Company for the Development of the São Francisco and Parnaíba Valleys. In 1952, the Bank of the Northeast (BNB) was created to help finance local economic activities. And, finally, in 1959, after another severe drought, the Superintendency for Development of the Northeast (SUDENE) was created with the objectives of promoting regional development and reducing the impact of the droughts (Hirschman 1963).

In order to achieve regional development, it was deemed necessary to create employment through industrialization, promote reorganization of the economy of the semiarid region to make it more resilient, and annex frontier regions to house excess population, especially the *cerrados* of western Bahia and the pre-Amazon region of Maranhão. Part of SUDENE's success derived from the prestige it achieved during the first years of its existence, between 1959 and 1964, when it was directed by the famous economist Celso Furtado, who, at the same time was the Planning Minister of Brazil. Fiscal incentives were defined to support productive activities in both the industrial and agro-ranching sectors, master plans were elaborated for development of the Northeast region and approved by the National Congress, and many studies were undertaken on the semiarid region and the Northeast region more generally. However, the country's political evolution led to the weakening of SUDENE and regional planning once power was recentralized in the federal government following the military takeover in March 1964. This situation did not change significantly after re-democratization of Brazil in 1985.

Another strong drought in 1970 again mobilized the political classes and the government. The federal government decided to reallocate resources to the Northeast region and created a National Integration Program (PIN), together with a Land Redistribution Program (PROTERRA). The solution to the problem of those dislocated by the drought, therefore, would be to facilitate their migration to Amazônia, on the one hand, and create new opportunities in the Northeast region, especially through irrigation, on the other. New roads were opened crisscrossing the Amazon region, including the famous Transamazon and Cuiabá–Santarém Highways, and agricultural *colonization* nuclei were established along their routes. This strategy, however, did not last for long as soils along the Trans-Amazon Highway proved to be too poor for permanent agricultural settlement purposes, and, starting in 1975, a new phase entailing the promotion of integrated rural development projects began, seeking to create better productive and living conditions for the rural workers and small farmers who were most adversely affected by the droughts. At the same time, both DNOCS and CODEVASF also stepped up their support for irrigation, although its benefits were initially limited (Hall 1978).

From this time, moreover, the World Bank began to support rural development interventions in the region. Yet another severe drought, from 1979 to 1983, led to re-evaluation of this approach and to the creation of the *Northeast Project*, which involved integrated rural development projects, complemented by other programs of rural sanitation, health, education, and land reform. Once again, a drought, in 1990, resulted in a reformulation of the approach, also with the World Bank's assistance, through which a program of decentralized, demand-driven rural community development projects was established by means of which local communities—and not the state governments—were to be primarily responsible for the planning and implementation of the actions involved. In addition, starting in the 1990s, efforts to manage the region's water resources were intensified with the creation of new institutions, the undertaking of physical investments in hydraulic infrastructure, and more proactive water supply and demand management.

In short, integrated water resource management also became a priority. During the implementation of this new process, politicians and technical specialists from the Northeast region visited water management experiences in the west of the United States, Mexico, and Spain, where they could witness first hand examples and extract lessons for application in the semiarid region of Brazil. Also at this time, the Water Resource Management Company of Ceará, COGERH, was established, and, in 2000, the federal government created the National Water Agency (ANA). Even though ANA covers the country as a whole, it has a special focus on the Northeast region and the problem of the droughts.

Meanwhile, significant political transformations had taken place both in Brazil as a whole and in the Northeast region. The military government, which began in 1964 and ended in 1985, was responsible for the large national development projects and programs briefly described above, including the Trans-Amazon highway, and thought that part of the solution for the Northeast drought problem was in the transfer of population to Amazônia. A civilian-led government, called the *New Republic*, came to power in the same year. And as noted above, in 1987, the traditional dominance of the large rural landowners in Ceará was replaced by emerging urban elites, and a new political class took control of the state government. Similar changes occurred in other northeastern states in subsequent years.

During this entire period, a National System of Civil Defense was officially responsible for recognizing drought situations and commanding emergency response actions. It was constituted by the Ministry of Interior (later MNI) in Brasília, SUDENE, which administered the system in the Northeast region, and the states. Special Drought Commissions also existed both at SUDENE and in the states. However, decisions were taken in a centralized way by the federal government and implemented by the states under SUDENE's supervision. This was the case, for example, in the severe drought of 1979–1983, when more than three million temporary jobs were created in the so-called emergency *work fronts* (Magalhães and Glantz 1992).

The year 1987 also witnessed a drought, but the government of the New Republic delayed reformulating its internal organization for drought management. Due to the urgent need to act, the states were allowed to define their own drought response measures. This led the Government of Ceará, for example, to innovate by taking measures that included greater transparency in the identification of the community works to be supported and in the selection of workers, with assistance of the beneficiaries through *community action groups*. This new approach practically eliminated the long-time resource diversion and elite capture problem and brought the beneficiaries closer to the definition and implementation of program activities. Other states adopted similar procedures and helped to improve the effectiveness of the emergency programs.

In 1995, the Aridas Project* (Magalhães et al. 1993; Projeto Áridas 1995) proposed a planning methodology for the development of the Northeast region containing a long-term vision and a focus on economic, social, and environmental sustainability, including climate-related crisis management, especially drought management, in an effort to bring these concerns together in a more effective and participatory manner.

1.7 Emergency Responses to the Drought

When drought impacts occur, the Government typically reacts in two complementary ways: one, with emergency actions, of an *assistencialista* nature, to reduce the suffering of the affected people; the other, of a more permanent character, with the objective of reducing their future vulnerability. Much of what has been described above refers to the second category, which is primarily concerned with water storage and improvement of its management so as to decrease the vulnerability of the regional population and economy. The following paragraphs describe the evolution of the emergency policies taken to alleviate people's suffering during the drought events themselves.

When a drought occurs, the affected population can be assisted in various ways, such as the following:

1. Through the emergency supply of water, the lack of which is a particularly serious problem for dispersed rural populations. During severe droughts, and especially when there are successive drought years, even the larger towns and cities need special water supply, such as through the use of tank trucks (*carros pipas*) or the drilling of wells.
2. Through the generation of jobs to ensure that people who become unemployed have access to a minimum income that permits them to acquire food supplies at least. Traditionally, this has occurred through the creation of temporary employment in what have been denominated as emergency *work fronts* (or *frentes de trabalho* in Portuguese), a type of Keynesian employment. Each worker enrolled in the work front received a salary. This strategy has evolved over time.

* The Aridas Project was developed during 1993 and 1994. It was headed by the Federal Ministry of Planning together with the Secretariats of Planning of all Northeast states and the participation of universities and civil society. The project built a methodology for planning sustainable development in the Northeast region, with a long-term view. It defined sustainability dimensions as social, economic, environmental, and political and established a seven-step planning process with the use of trend and desired scenarios and the inclusion of risk of climate variations (variability and change) presently and in the future. The Aridas Project influenced sectoral policies in Brazil, such as the water policy and state development planning in the Northeast region.

Two aspects of this type of assistance have been particularly discussed over the years: what types of civil works to undertake and how to link them with medium- and long-term regional and local development objectives, and what salary to pay, with some arguing that this should be a minimum value so as not to compete with possible productive jobs.

3. Through crop insurance or some other form of compensation for farmers who lose their harvests because of the drought.
4. Through the sale of grain at subsidized prices to feed livestock after pastures have been decimated.

Guaranteeing water to the dispersed populations has been a common response during periods of droughts. This has been done for many years, and continues to be so, through the use of tank trucks (*carros pipas*) that bring water from existing sources, generally reservoirs or wells, to the isolated communities. In 2014, for example, 8000 tank trucks distributed water throughout the semiarid region, counting only those financed by the federal government, together with thousands of others financed by the states. In cases of annual or moderate droughts, *carros pipas* need to attend only to parts of the rural zone; but in multiyear and more severe droughts, such as the most recent one, which is still ongoing as this book goes to press, even larger cities need to be supplied by *carros pipas* or other means, such as the construction of emergency aqueducts.

The work fronts existed throughout the twentieth century. In some cases, larger public works were built requiring the dislocation of workers, who were separated from their families for some time. This often resulted in family break-ups and even had political repercussions. On other occasions, as in 1983, smaller works were constructed closer to the communities, thus requiring fewer dislocations. More than three million emergency jobs were created in the Northeast region in that year. The salary was very low, only a fraction of the legal minimum wage, but various family members could be employed. In 1987, a new experiment was tried in states like Ceará, in which only one person per family could be employed, but the works were of interest to the local community and the legal minimum wage was paid. As a result, children were still able to go to school. Different types of activities were also supported, including local health agents, which led to the establishment of a broader national health agents program.

Crop insurance and the sale of subsidized corn for cattle raisers are also actions that have been practiced during recent droughts with some success. Small farmers who live in municipalities that lose more than 50% of their harvest receive a certain quantity of government resources to compensate for their expenses. And the government has sold subsidized corn to help feed drought-affected livestock. This did not prevent many heads of cattle dying from thirst and hunger, even in more recent drought years, although the number was lower than in earlier droughts.

1.8 The 2010–2015 Drought

Between 2010 and 2015, the Northeast region has experienced another period of irregular rainfall. The year 2010 was dry, with the loss of harvests. However, the reservoirs were full due to more plentiful rainfall during the two immediately preceding years, so that there was not a major problem with respect to water supply. The rainfall level in 2011 was about average, and even though there was no significant additional accumulation of stored water, agricultural production was abundant. Thus, people were able to cope with the income generated from these activities. However, starting in 2012 and lasting through the present (i.e., 2015), the region has been afflicted by another severe multiyear drought.

Thus, 2015 was the fourth consecutive year of drought. This is very serious because, in addition to the loss of agricultural output, small, medium, and even some large reservoirs have dried up, and the tank trucks need to bring water from ever more distant localities. Many rural towns that are municipal or district seats are now without local water supplies and lack alternative sources. Ensuring an adequate supply of water, therefore, has become the greatest challenge associated with the present drought. As this book is being written, there is considerable concern about 2016, which is expected to be a strong El Niño year. This normally means drought in the Northeast region. If this occurs, the present drought will become an even more extreme one because the water sources that supply the region are already depleted. Even the largest cities will be threatened with lack of water.

Over the past decade or so, however, there have been some important developments with respect to the income generation component of the emergency policies. With the World Bank's support and the leadership of the Ministry of Social Development, Brazil, has established a system of conditional cash transfers to the poorest families, although this program does not apply to the semiarid region of the Northeast region alone. At present, some 14 million families throughout Brazil receive income transfers from the federal government through the *Bolsa Família* program. This number includes the poor families of the northeastern region traditionally affected by drought and whose members have provided their labor to the emergency *work fronts*. In short, while the purpose of the *frentes de trabalho* was to provide a minimum income for workers and their families, this objective is now being met by *Bolsa Família* during the present drought. As a result, despite the drought's severity, there was no longer a need for the emergency fronts. More than half the benefits of *Bolsa Família* go to families in the Northeast region. Even though this region accounts for only 28% of the national population, it contains the largest concentration of rural poverty. Consequently, some 8.6 million families in the region receive resource transfers on a monthly basis from this program. To maintain their eligibility for these payments, families are required to keep their children in school and meet certain obligations with respect to use of the national health system, including vaccinations and pre-natal care (Magalhães and Martins 2011; MDS 2015).

1.9 The Droughts and Climate Change: An Emerging Threat

According to the Intergovernmental Panel on Climate Change (IPCC 2014), some of the largest impacts of global climate change will occur in the semiarid regions of the planet, including Northeast Brazil. These results are confirmed by the Brazilian Climate Change Panel (PBMC 2013). In accordance with various climate models, it is probable that the frequency and intensity of droughts in the Northeast region will increase, together with the duration of the drought periods, currently being experienced in certain parts of Brazil. Global temperatures are expected to rise at least 2°C by the end of the century. There will be greater evapotranspiration and thus less humidity in the soils. Several models indicate that important river basins in the Northeast region, including the São Francisco, will have reduced water flow. With stronger droughts, lower humidity, and less water, impacts on agriculture, the economy, and living conditions will also be greater. In addition to the historical and traditional challenges faced in addressing drought impacts in the region, accordingly, new ones are presented by the prospect of more frequent and intense dry years.

In the case of the Northeast region, many of the future impacts of climate change are likely to resemble the effects of climate variability associated with extreme droughts, such as the one between 2010 and 2015. Just as in the case of the present drought, with increasing climate change the most vulnerable sectors will be rain-fed agriculture and water supply for human consumption and other uses. But other sectors such as irrigated agriculture and hydroelectric energy production will also need to adapt to the new global and regional climate situation. The better Brazil prepares to confront the drought problem and promote adaptation to the prevailing conditions in the semiarid region, the easier it will be to respond to future climate changes.

1.10 The Need for Proactive Responses

The Northeast region has a long history of confronting droughts. This notwithstanding, droughts continue to involve an element of surprise. Despite the evolution and improvements in the way the federal and state governments have responded in recent years, some improvisation has tended to occur each time a new drought happens and the society needs to react. Now, new elements are added to the evolving scenario due to increased environmental vulnerability resulting from processes of land degradation and desertification due to increasingly unsustainable use of soils and water, as well as to the changing climate, which is already starting to be felt in the region.

Until now, government responses have been reactive: what is to be done is only decided once a new drought begins. A reactive response not only refers to the emergency actions to be taken. The history of droughts in the Northeast region reveals that policy makers have been concerned both with emergency measures and actions to reduce future vulnerability. But these decisions have generally always been made as a response to each new drought. This was the case with the establishment of the *hydraulic phase* after the drought of 1877–1879, and with the creation of DNOCS, BNB, SUDENE, and special programs (e.g., PIN and PROTERRA) following subsequent ones. The emergency work fronts reflected the specific conditions of each drought, including the prevailing political structure and the priority attributed to it, almost always as a function of the level of drought calamity.

This situation needs to change. Brazil has to move in the direction of a proactive policy to confront future droughts and their impacts. It now appears that political and technical conditions exist for this to occur. The process to improve the quality of response to droughts in Brazil and to change the policy paradigm from one that has been traditionally reactive to another that is more proactive is described in more detail in the following chapters.

2

Crisis, Opportunity, and Leadership

Eduardo Sávio P.R. Martins, Francisco José Coelho Teixeira, João Gilberto Lotufo Conejo, José Machado, and Antonio Divino Moura

CONTENTS

2.1 The Crisis 19
2.2 The Opportunity 22
2.3 Leadership 23

2.1 The Crisis

As observed in Chapter 1, the recent hydrological crisis in Brazil has affected not only its northeastern region, commonly subject to recurrent droughts, but also the large metropolitan areas of São Paulo, Belo Horizonte, and Rio de Janeiro. The present water crisis in the Northeast region is the result of a multiyear drought, which has thus far lasted from 2010 to 2015. Although most of this book focuses on the Northeast region, it is also useful to understand the current drought situation in other parts of Brazil, as it shows the vulnerability of even predominantly wetter regions of the country to extreme prolonged droughts.

In the case of the metropolitan Southeast region, the extent of the drought was also particularly severe, both because of its longer duration (2014–2015) and its above-average intensity, representing the most significant one observed over the past 80 years. The graph in Figure 2.1 illustrates the severity of the recent drought in the southeastern region during which monthly water discharges into the principal reservoirs of the Cantareira system in metropolitan São Paulo reached levels significantly lower than their historical averages and during the most critical prior year of record (1953).

Government responses to the past and present drought crises have a number of common characteristics, including, for example, problems in communicating the seriousness of the situation to the population; difficulty by decision makers to take unpopular measures, such as rationing; and lack of a long-term perspective in terms of drought management, as noted in Chapter 1. An analysis of these crises also reveals that political considerations, especially in an election year (such as 2014), can hinder or delay assessment of

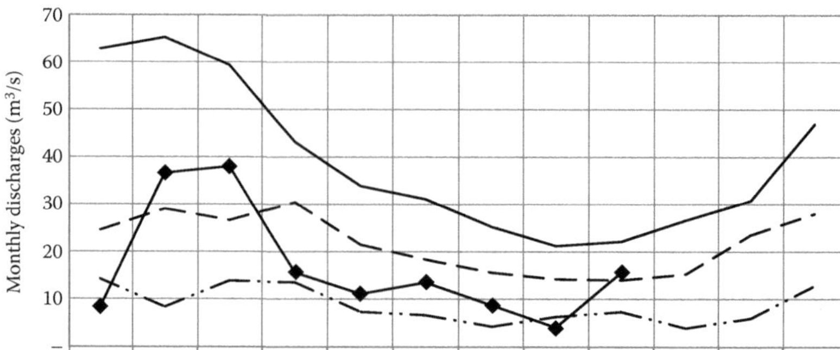

	Jan	Feb	Mar	Apr	May	Jun	Jul	Aug	Sep	Oct	Nov	Dec
—— Average	62.84	65.41	59.52	43.19	33.90	30.98	25.18	21.27	22.18	26.81	30.75	46.97
– – 1953	24.53	29.14	26.75	30.44	21.46	18.46	15.62	14.19	14.07	15.44	23.72	27.93
—·· 2014	14.32	8.47	13.77	13.46	7.26	6.62	4.17	6.28	7.25	3.96	6.04	12.77
—◆— 2015	8.50	36.55	38.14	15.59	11.19	13.63	8.74	4.09	15.50			

FIGURE 2.1
Average monthly discharges (m^3/s) into the principal reservoirs of the Cantareira system; historical average, recent years (2014/2015) and the previous most critical year (1953).

the gravity of the drought situation and adversely affect decision making for mitigating its effects, again particularly when unpopular actions may be required (Revista Época, 1985).

In the case of São Paulo, the 2014–2015 crisis contained similarities with previous ones, especially that of 1985, when the state government relied on the Cantareira system to respond to the problem. During the recent crisis, the state sought, as an emergency measure, to use the *dead storage* of the Cantareira system reservoirs by pumping, reinforced by other sources to meet demand. However, ultimate resolution of the crisis will depend on the robustness of the 2016 rainy season. This notwithstanding, the strategies adopted in São Paulo in response to the current crisis represented good policies, including for example, reduced pressures on the water supply system, diversification of water sources for the metropolitan area, and establishment of a bonus program* for those who economized on water use, generating notable impacts in terms of reduced demand.

During the same period, drought conditions in the Southeast region also affected the Belo Horizonte and Rio de Janeiro metropolitan areas. The crisis in the Southeast region, in addition to being hydrological, was also a human. As such, it required a long-term vision and an effective policy to combat drought effects. It also required integration of drought management with other policies that seek to provide adequate infrastructure and water supply in response to the accelerated growth of demand in the metropolitan

* The São Paulo water distribution company (SABESP) determines a bonus of 10%, 20%, and 30% for those who reduce water consumption, respectively, in the ranges 10%–15%, 15%–20%, and over 20%. Also, a contingency fee is charged to customers whose monthly consumption exceeds the average calculated from February 2013 to January 2014.

areas and to make their occupation more sustainable and, thereby, to reduce the stresses generated by high and unplanned population concentrations. To a certain extent, moreover, these problems are observed throughout Brazil and, thus, are challenges that are national in scope.

Back to the Northeast region, the heavy investments in water-related infrastructure in the previous decades were not sufficient to provide a definitive and effective response to drought impacts especially to those associated with the multiyear drought of 2010–2015. It is now evident that the solution not only requires improved infrastructure, but also the identification of vulnerabilities and planning for specific contingencies in each sector and for each specific water management system (e.g., contingency plans for urban water supply systems for the metropolitan regions, and for each reservoir's system). In other words, better infrastructure and resource management are both required to guarantee water security.

The initiative to develop and implement the Northeast Drought Monitor, which is described in Chapters 4 and 11, occurred during this current extended drought in the region. During this period, various problems, some of which were similar to those experienced in the southeastern crisis, made a concerted and adequate response to the effects of this drought difficult, among which were the following:

- Those of an institutional nature (e.g., the lack of qualified human resources)
- Those of a political character (e.g., lack of transparency* and social communication)
- Those of infrastructure vulnerability (e.g., insufficiency of the reservoirs built to supply the ever-growing demand for water)
- Those of a logistical nature (e.g., decentralized storage of corn for distribution)
- Those due to the fragility of institutional articulation among different levels of public administration (for instance, drought mitigation programs implemented in an uncoordinated way)

Despite the problems observed during the 2010–2015 drought in the Northeast region, social protection and drought alleviation programs have played an important role in terms of mitigating its potential adverse social impacts, such as increased outmigration, looting, and hunger, as indicated in Chapter 1. Until, at least mid-2015, these dramatic effects have not been experienced in the region, but, considering the elevated risk that the drought

* A time of political transition can also affect a drought response in terms of public finance. In 2013, for example, the mayor who had assumed control of a municipality in the semiarid region that was visited by one of the authors of this chapter did not have access to the available resources to respond to the drought in 2012–2013 in part because his predecessor had withheld pertinent information (Magalhães and Martins, 2013).

will persist in 2016, together with the severe current national economic crisis, they could still recur in the future.

The magnitude of the crises provoked by the droughts of 2010–2015 in the Northeast region and of 2014–2015 in the southeastern region, however, has meant that the question of water resource management associated with water security has entered the national political agenda, creating an opportunity to better structure and prepare the country to confront future droughts and dry periods. What is needed now is to take advantage of the lessons of the past.

2.2 The Opportunity

The historical perspective presented in Chapter 1 reveals that successive Brazilian governments have met the recurrent drought crises in the Northeast region primarily through repetitive reactive measures. Reactive measures do not necessarily mean *emergency* ones, but they refer also to longer-term actions, whether permanent or not, taken as a response to a specific drought emergency. On some occasions, as in the early 1960s, decision makers resolved that socioeconomic development measures through the use of planning tools and government incentives should prevail over this traditional emergency response logic. However, these proved to be fleeting moments that subsequently faded as the result of political changes. Chapter 4 describes the current effort for drought monitoring, a key element to accomplish a paradigm shift in drought management: from a crisis management (or reactive) one to a risk management one, or by proactively treating the causes and not just the symptoms by having mechanisms in place to better anticipate drought events and guide relief measures more efficiently, objectively, and effectively.

The 2010–2015 drought in the Northeast region reached such a proportion that it again caused the principal decision makers in the federal and affected state governments, as well as in society as a whole, to consider the need to act on the basis of new assumptions. The constant warnings issued by the international scientific community about climate change and its impacts, and in particular, about the likelihood that these changes would make droughts more frequent, prolonged, and severe, led even the most skeptical individual to worry about the improper treatment given to droughts in the past, thereby opening a new window of opportunity to rethink public policies in relation to them.

Once again, leaders both within and outside government called attention to the need for a better structuring of drought policy for the region, such that it would promote greater articulation among the individual responses of the various institutions involved at different administrative levels (municipal, state, and federal). They also drew attention to the need to introduce a vision of the future into the planning for drought response, moving from a current crisis management approach to a longer-term risk management one, or, in other words, to a more permanent and proactive treatment of the problem. Even so, they recognized that emergency measures would still be necessary in crisis situations. Moreover, instead of taking decisions in the heat of the situation, they should be previously foreseen and included in a coordinated planning process and management system to avoid resource allocation occurring at random or in a fragmented and irrational way.

The discussion regarding the droughts and their impacts has reached the highest levels of the federal executive as well as the state legislatures and national congress. The underlying water crisis is now perceived nationally as a central problem, and the topic of water resource management, which had been striving to gain ground since a new National Water Law was promulgated in 1997, has finally become a key element on the political agenda, in the process exposing the institutional fragilities to deal with situations of hydrological uncertainty and water deficits.

This growing awareness at the highest political level about the seriousness of the climate change challenge and its hydrological implications, together with the aforementioned institutional weaknesses, have led to a golden opportunity for a paradigm change in the Brazilian government's actions to deal with critical drought events. The unprecedented climate and hydrological crisis, given its breadth and depth (and since it is no longer restricted to the Northeast region, but now also affects the most developed parts of the country), clearly strengthens the rationale for elaborating a National Drought Policy with risk, instead of crisis, management as its guiding principle and overcoming the reactive and emergency character of government actions of the past.

2.3 Leadership

The policy responsibility for natural disasters, such as droughts in Brazil, lies, constitutionally, with the federal government, which acts in conjunction with the affected states. Within the federal government, the principal responsible institution is the Ministry of National Integration (MNI), where the National Civil Defense Secretariat is located and which is always mobilized to undertake emergency actions in response to disasters.

MNI was created in 1999 with a wide-ranging mandate that includes (1) formulation and implementation of the National Policy of Regional Development and National Irrigation Policy; (2) support for the formulation and implementation of the National Civil Defense and Protection Plan; and (3) support for the construction, operation, maintenance, and rehabilitation of major water infrastructure. Consequently, it

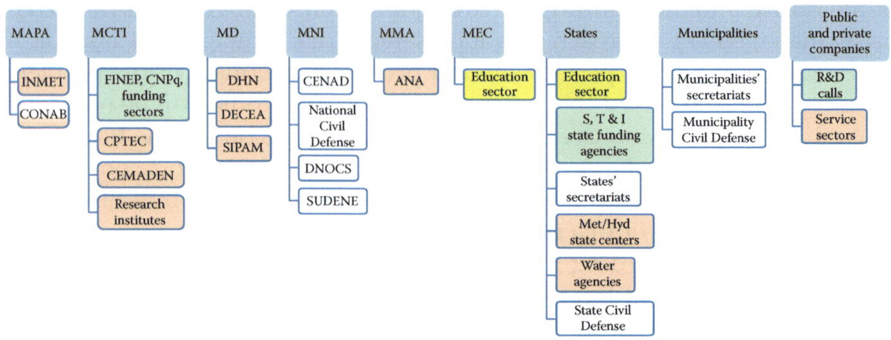

FIGURE 2.2
Institutions involved in drought management in Brazil, with those associated with training identified in yellow; with financing of studies and research in green; the centers responsible for monitoring, forecasting, processing, and disseminating meteorological, hydrological, or agricultural information in pink; and the users of these products and information in white. MAPA, Ministry of Agriculture, Livestock and Food Supply; MCTI, Ministry of Science, Technology and Innovation; MD, Ministry of Defense; MNI, Ministry of National Integration; MMA, Ministry of the Environment; MEC, Ministry of Education; S, T & I, Science, Technology and Innovation; INMET, Brazilian National Institute of Meteorology; CONAB, National Food Supply Agency; FINEP, Funding Authority for Studies and Projects; CNPq, National Council for Scientific and Technological Development; CPTEC, Brazilian Center for Weather Forecasting; CEMADEN, National Centre for Monitoring and Warnings of Natural Disasters; DHN, Directorate of Hydrography and Navigation; DECEA, Department of Airspace Control; SIPAM, Amazon Protection System; CENAD, National Center for Risk and Disasters Management; DNOCS, National Department of Works Against the Drought; SUDENE, Superintendency for Development of the Northeast; ANA, National Water Agency; R&D, Research and Development.

plays a key role in elaborating and executing strategies and plans to alleviate the effects of droughts. However, many other government agencies are also involved (as illustrated in Figure 2.2).

Despite this attribution of permanent responsibility, certain specificity is involved in the response to each drought. To orchestrate the emergency actions of an *assistential* nature, the most recent practice (i.e., since the year 2000) has been to install drought committees, led from within the President's office at the federal level, and by committees or sectoral secretariats at the state level. These committees are temporary organizations and, thus, tend to have *ad hoc* institutional arrangements and are often criticized for not being able to respond rapidly with comprehensive and integrated actions. Planning and integration across institutions have constituted one of the main difficulties in relation to government responses to the droughts in Brazil.

The government response at various levels to confront and mitigate the effects of the prolonged severe drought of 2010–2015 in the Northeast region has been of an unprecedented intensity in the context of crisis management. The federal government created the Drought Observatory, linked to the Chief of Staff of the Presidency of the Republic (*Casa Civil*), with the principal aim of giving greater visibility to the investments made in response to this event. MNI, in turn, established an Integrated Committee to Combat Drought in the semiarid region in 2012 to coordinate government actions to address the drought in the Northeast region and Minas Gerais. However, neither the Drought Observatory nor the Integrated Committee included planning (short-, medium-, or long-term) among its responsibilities.

Responses, in particular, by the federal government, have included, in addition to the traditional measures taken in the past (such as the use of tank trucks, well drilling, opening of emergency credit lines, and renegotiation of farmers' debts), other actions such as the *Bolsa Estiagem** (or *drought allowance*) and *Garantia-Safra*† (*harvest guarantee*). These emergency measures, which together with *Bolsa Família*, which is permanent in character, have provided vital assistance to family farmers, as already mentioned in Chapter 1. In addition, each drought-affected municipality received, as a donation from the federal government, a fleet of vehicles including a digger, a grader, a bulldozer, and a dump truck for emergency civil works together with a water tank truck. Altogether, the federal government has spent a considerable amount of resources for emergency actions during the crisis, the great majority of which coming from MNI through the National Civil Defense Secretariat. When these are added to other structural actions under implementation, such as permanent large-scale hydrological infrastructure works (among them, the project to integrate the São Francisco River with the river basins in the northern part of the Northeast region, which will allow the transposition of water from the former to the latter, benefiting the states of Ceará, Rio Grande do Norte, Paraíba, and Pernambuco), federal government expenditures have reached some $US10 billion.

As observed above, in 2012, the nature and severity of the drought in the Northeast region revived the need to truly shift to a new paradigm for drought management. Thus, in parallel to crisis management, which followed its traditional course, several new initiatives were undertaken starting in that year. Among the most noteworthy of these was the creation of a working group by MNI and its associated agencies, which was tasked with studying the drought and proposing measures to address it.

This working group was established in the context of a global discussion about national drought policies. The Ministry had received an invitation from the World Meteorological Organization (WMO), in conjunction with the United Nations Convention to Combat Desertification (UNCCD) and the United Nations Food and Agriculture Organization (FAO) to participate in

---

* *Bolsa Estiagem* is a temporary cash transfer program to poor families affected by droughts.

† The goal of this program is to provide income security to farmers in case of droughts or floods. This program was established by the federal government as part of the National Program for Strengthening of Family Farming (PRONAF). It is given to the farmers of the semiarid regions of Brazil who lost their crops due to droughts or excess rains. The crops covered by this program are beans, corn, rice, cassava, and cotton.

and co-sponsor a High Level Meeting on National Drought Policy (HMNDP), in Geneva, Switzerland, in March 2013 (WMO 2013). A Brazilian Government mission, led by the then Secretary of Water Infrastructure of MNI and composed of specialists from several federal institutions including the National Institute of Meteorology (INMET) and the government of Ceará, attended this event and presented the Brazilian experience based on a report prepared by the Working Group and inputs from Ceará. The Secretary of Water Infrastructure later became the Minister of MNI, thus assuring important support for the elaboration of a National Drought Policy for Brazil.

During the Geneva meeting, the Brazilian delegation heard from technical specialists and representatives of other governments that, without a coordinated national policy with respect to droughts, countries would continue to respond with reactive a *posteriori* methods (i.e., crisis management). A coordinated national policy includes effective early warning and dissemination systems, to transmit timely information to decision makers, as well as effective procedures to evaluate drought effects, proactive risk management measures, prevention plans to increase the capacity to confront droughts, and effective response programs in emergency situations aimed at reducing drought impacts. The MNI, convinced that this was the path to be adopted and complementary to the efforts of the Ministry's Working Group, requested technical assistance from the World Bank to (1) structure and facilitate a more permanent institutional approach and response to drought, and to improve integration within and between federal and state institutions and (2) develop concrete drought monitoring tools and preparedness plans/protocols for specific sectors.

Through a collaborative effort between MNI, ANA, INMET, the Center for Management and Strategic Studies (CGEE), the World Bank, and the governments of the Northeast states (especially the Ceará Foundation for Meteorology and Water Resources—FUNCEME), a series of planning and convening discussions took place in the later part of 2013 and early 2014 concerning droughts in the Northeast region. These discussions involved all three levels of public administration in Brazil, together with specialists, and representatives of civil society in a highly participatory process. The impacts of the 2012–2013 drought were assessed, as were the response measures taken to address them. The extensive past experience with drought management in the Northeast region was also reviewed, and the experiences of other countries that have practiced risk management as part of their drought planning activities, especially the United States, Mexico, Spain, and Australia, were presented.

This process resulted in a proposal to restructure the drought policy for the Northeast region with the intention to later extend it to the rest of the country, based on three pillars, which are presented in greater detail in Chapter 3.

3

A Framework and Convening Power

Erwin De Nys, Nathan L. Engle, and Carmen Molejón Quintana

CONTENTS

3.1 The Paradigm Shift..28
3.2 The Convening Power..29
3.3 The Three Pillars of Drought Preparedness..32
3.4 The First Steps of Change..33
3.5 Conclusion..34

As indicated in Chapters 1 and 2, the recent prolonged drought in the Northeast and Southeast Brazil have called attention to the need for greater coordination across the different sectors and levels of government involved. Technical solutions to strengthen the resilience of populations, animals, and crops to recurrent droughts are plentiful. Specialized institutions with the responsibility to plan the socioeconomic development of regions afflicted by such droughts also exist. However, many in Brazil perceive insufficient articulation of actions and investments among those institutions as the main bottleneck responsible for delays in the emergency response to and preparedness planning for the most recent drought.

This chapter elaborates on the underpinning framework for the paradigm shift that Brazil has been aiming to bring about in recent years. It explains the desired change from the traditional crisis management, or reactive, approach, to a risk-based drought preparedness one that is iterative and predicated on greater proactivity. This chapter describes the guiding structure that supports this approach, which consists of three equally important pillars. First is the capacity to forecast a drought occurrence and to monitor and communicate its evolution over space and time. Second is the ability to assess and track the vulnerability and risks of populations and ecosystems to droughts and their impacts. Third are the necessary actions and investments to mitigate these impacts together with their underlying supporting policies, plans, and decision-making processes.

Finally, the chapter provides insights into the challenges of institutional coordination for drought management in the Brazilian context. The paradigm shift must occur collectively among the numerous institutions that have had traditional crisis management engrained in their disparate approaches to drought policy and planning. This chapter illustrates the concrete first steps made by Brazil to shift this paradigm, with the convening power and support of partners at the national and international levels, and identifies a few areas that may define the way forward for Brazil in the coming years.

3.1 The Paradigm Shift

As a starting point, it is useful to summarize the cycle of disaster management, which is applicable to droughts, floods, and earthquakes, among other disasters (Figure 3.1). This cycle has two distinct parts. The first is traditional crisis management. Many governments and donors have typically followed this reactive approach. It consists of a series of actions subsequent

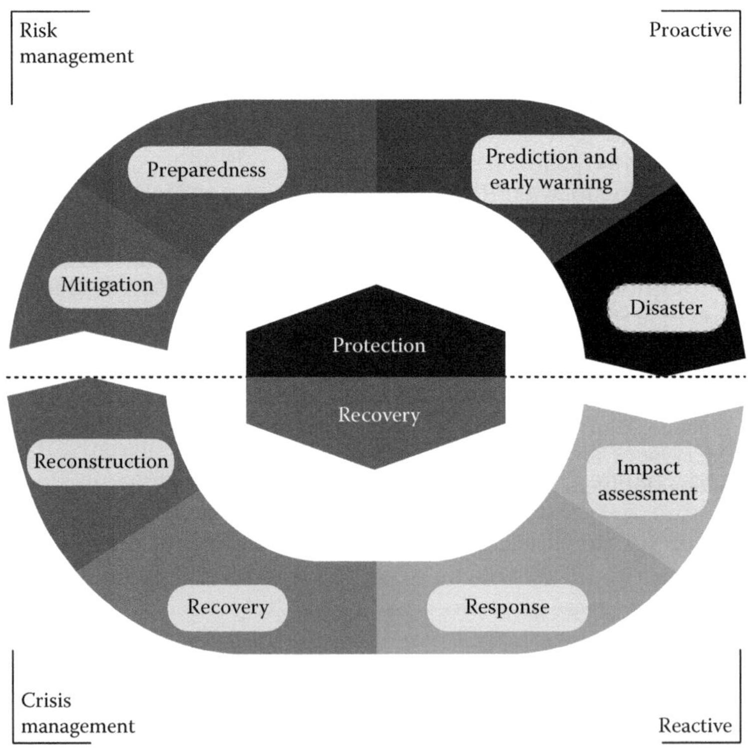

FIGURE 3.1
The cycle of disaster risk management. The typical reactive and crisis management emphasis of droughts is noted on the bottom half of the figure, whereas the paradigm shift pursued by Brazil toward more proactive drought preparedness and risk management is noted in the top half of the figure. (From National Drought Mitigation Center, University of Nebraska–Lincoln, Lincoln, NE.)

to a disaster, including an assessment of its impacts, response, recovery, and reconstruction actions to restore the affected locality or region to a pre-disaster state. Historically, governments have given less attention to the second part of the cycle, risk management. Risk management includes the proactive actions that precede the disaster, and which aim to avoid or reduce future impacts. Such actions include early warning and monitoring, planning, mitigation, and the development of risk-based national drought management policies.

Crisis management only addresses the effects and impacts of droughts that occur as a direct or indirect consequence of drought events. Risk management, on the other hand, focuses on identifying vulnerabilities and systematically and iteratively implementing measures to lessen the potential impacts associated with future droughts.

In Brazil, the federal government's approach toward drought has historically relied on the reactive crisis management part of the cycle. As highlighted in Chapter 2, the *Casa Civil*, a federal office analogous to a presidential chief of staff, has traditionally had a strong bias toward emergency relief and public works, often relying on *ad hoc* committees to deploy response and recovery actions. However, the severity of the impacts of the present (i.e., 2010–2015) drought has highlighted the inadequacy of this approach, often characterized by slow response and poor coordination within and between the federal and state institutions involved. In response, several leading ministries and agencies have expressed the need for a more proactive approach and a more stable institutional framework to deal with this phenomenon, as noted in Chapters 1 and 2.

3.2 The Convening Power

At the international level, growing concern about the spiraling impacts of droughts has led several key organizations to promote a paradigm shift from crisis to risk management through the development of national drought management policies. Most notably, as indicated in Chapter 2, the World Meteorological Organization (WMO), the Secretariat of the United Nations Convention to Combat Desertification (UNCCD), and the United Nations Food and Agriculture Organization (FAO) organized a first of its kind High-Level Meeting on National Drought Policy (HMNDP) in Geneva, Switzerland, during March 11–15, 2013. It counted on the participation of experts and policy makers from 87 countries, as well as other UN agencies, international and regional organizations (Wilhite 2014b). The result was a mutually agreed declaration recognizing the needs to build drought resilience and national drought policies to help facilitate this process. Regional capacity-building workshops have followed the HMNDP with the aim of helping drought-prone countries formulate and adopt risk-based national drought management policies (Tsegai and Ardakanian 2014). At the HMNDP, one of the countries most determined to develop more proactive drought management policies was Brazil, represented by its then National Secretary of Water Infrastructure, under the Minister of National Integration.

The Secretary of Water Infrastructure had called upon the technical expertise and convening power of several national and international organizations to help prepare for the HMNDP. At the national level, these included the National Water Agency (ANA), the National Meteorological Institute (INMET), the Brazilian Agricultural Research Corporation (EMBRAPA), the National Center for Monitoring and Early Warning of Natural Disasters (CEMADEN), and the Center for Strategic Studies and Management (CGEE). ANA, in particular, plays an important technical leadership role with respect to droughts through its responsibilities such as the monitoring of water supply and availability at the national level, as well as the management of federal river basins (i.e., those watersheds that cross state and international boundaries).

As mentioned in Chapter 2, the World Bank was also brought in to help the Ministry develop a program of technical collaboration on drought preparedness and climate resilience (referred to throughout this book as the "technical collaboration program"). The World Bank is a longstanding development partner of Brazil, particularly in the Northeast region, having supported the federal and state governments with water resource management, water supply and sanitation, irrigation, regional and rural development, and disaster risk management (DRM) for many years. Other World Bank-financed projects related to drought mitigation in the Northeast region have focused on community and regional development, income diversification, and climate-smart agriculture, among other priorities.

The main objective of the technical collaboration program for the Ministry of National Integration (MNI) was to help stakeholders in Brazil (both at the national level and in the Northeast region) to develop and institutionalize proactive approaches to recurrent drought events, with the ancillary benefit of developing tools, frameworks, processes, and exchange platforms from which other countries could also learn and foster innovation around this topic. The World Bank played an important role in developing and organizing a conceptual framework for the desired paradigm shift by convening national and international expertise, including from the state of Ceará (FUNCEME, Ceará Foundation for Meteorology and Water Resources), Mexico (CONAGUA, the National Water Commission), Spain (several local government and academic experts), and the United States (the National Drought Mitigation Center—NDMC—and several other academic experts). Finally, the World Bank, together with MNI and its partners, was fundamental in involving all nine states of the Northeast region in this program (Figure 3.2). Much of the experience described in the remaining chapters of this book is based on the process and outcomes of this technical collaboration program.

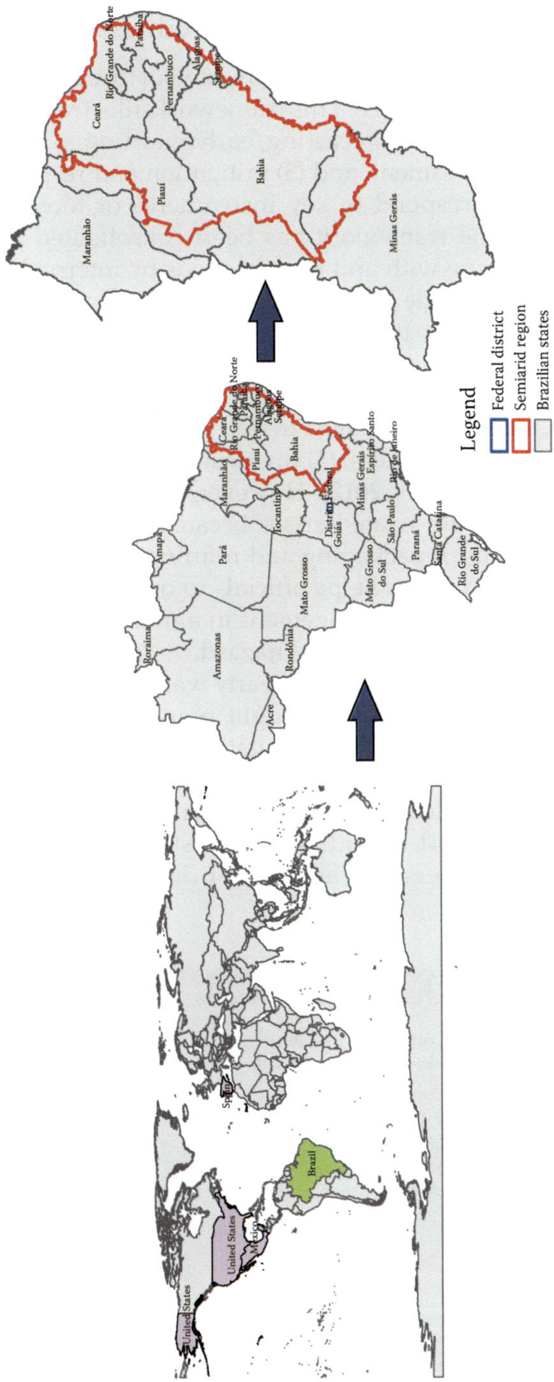

FIGURE 3.2
Map of Northeast Brazil and the world depicting the international partners involved in the drought preparedness and climate resilience program.

3.3 The Three Pillars of Drought Preparedness

The guiding framework for this program consisted of three pillars of drought preparedness, of equal importance. This framework, illustrated in Figure 3.3, consists of (1) monitoring and forecasting/early warning, (2) vulnerability/resilience and impact assessment, and (3) mitigation and response planning and measures. They correspond to key instruments or tools of a drought preparedness policy. The framework has been consolidated from various sources and from dialogue with and presentations by international drought experts (e.g., Bazza 2001; Engle 2012; Wilhite et al. 2014a).

One initial challenge that the program encountered when introducing partners and stakeholders to this framework was how these "three pillars" were inherently different from DRM frameworks, such as the World Bank's (World Bank 2012), which defines DRM in five pillars—(1) risk identification, (2) risk reduction, (3) preparedness, (4) financial protection, and (5) resilient disaster recovery (World Bank 2012). However, this general framework is not well-adapted to drought-related crises because of their slow onset and gradual occurrence. Thus, repackaging and refining this model to the three pillars of drought preparedness helps officials to operationalize DRM principles for purposes of drought management in a manner that most directly applies to this particular type of natural hazard.

The first pillar refers to monitoring and early warning systems that are the foundation of the other elements of drought preparedness. It includes the monitoring of relevant indicators (precipitation, temperature, evapotranspiration, seasonal weather forecasts, soil moisture, stream flow, groundwater, reservoir and lake levels, etc.) and the use of appropriate indices through a coordinated effort by individuals, institutions, and information systems. It requires the integrated analysis of data with tools that can be

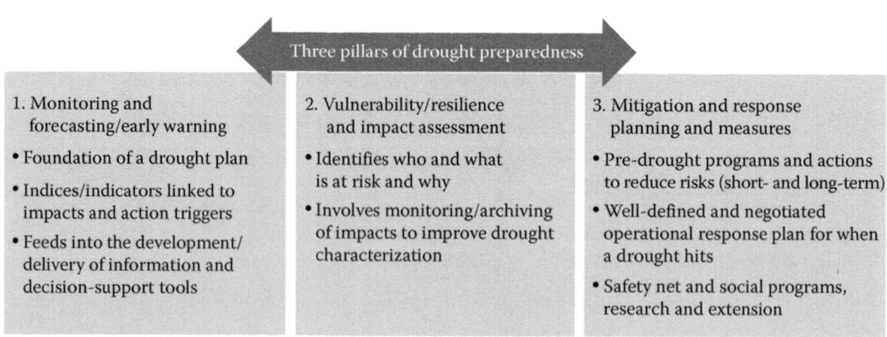

FIGURE 3.3
The "three pillars of drought preparedness" that serve as the guiding framework for the technical collaboration program to the MNI to support a paradigm shift away from reactive crisis management to more proactive approaches to the management of drought events.

used by decision makers to detect droughts and improve preparation for and response to them through associated actions and triggers.

The second pillar, vulnerability/resilience and impact assessment, encourages stakeholders to engage in a dialogue on risks before the occurrence of droughts so that priorities can be identified and negotiated based on "who" (i.e., which stakeholders) and "what" (which economic sectors) should be involved in drought preparedness and response. In addition, indicators and impact-reporting procedures established through these assessments can help improve the timing and expediency of planning and management once a drought event starts. Tracking impacts can also provide critical information for the monitoring and evaluation of the socioeconomic benefits and costs of drought preparedness so that communities can work to strengthen their capacity and resilience.

The third pillar relates to mitigation and response planning to develop proactive measures to increase a community's coping capacity as well as response measures aligned with the principles of risk reduction. This pillar includes an operational drought response plan having pre-negotiated triggers and actions for when and how different sectors should respond to mitigate drought impacts. Importantly, based on assessments under the second pillar and linked with the monitoring/early warning systems in the first, short- and long-term structural measures are intended to tackle societal vulnerabilities (Tsegai and Ardakanian 2014b). In this sense, having elements of the plan that can be implemented in "nondrought" times is important for building long-term drought preparedness and resilience.

The interaction among the three pillars indicates that drought planning should be viewed as an ongoing process (Wilhite et al. 2005), which is predicated on a solid monitoring and early warning system, linked to the evaluation of a community's or region's vulnerability and resilience, and related to policy and management action triggers and long-term investment decisions.

3.4 The First Steps of Change

Stemming from the HMNDP-related efforts, at the international level, generic steps or guidelines have been developed so that nations can apply the overarching principles of a National Drought Policy aimed at risk reduction (Wilhite 2014b). This policy would be carried out at the subnational level through the implementation of Drought Preparedness Plans that follow the framework and principles of the National Drought Policy. This process requires political will at the highest level and a coordinated approach within and between different levels of government as well as among the other stakeholders that must be engaged in the policy development process.

In Brazil, given its continental size, the technical collaboration program's proposition was to develop the key tools or instruments of a proactive national drought management policy, starting, on a pilot basis, at the level of a specific region, the Northeast, and subsequently expanding it to cover the country as a whole. The program helped to develop two instruments that were seen as fundamental to move toward a policy based on the principles of drought preparedness and the three pillars framework briefly described above: a Drought Monitor for the entire Northeast and drought preparedness plans for specific locations within the region.

In its most visible form, the Northeast Drought Monitor is a map, updated monthly, which describes the current drought stage conditions across the region, designed on the basis of several meteorological and hydrological indicators. The indicators are weighted to produce a composite five-stage drought severity index and thus add greater nuance, objectivity, and consistency to the definition of drought conditions across the region. Beyond the map, the Monitor is an organizational arrangement of people, institutions, and processes that are as important as the drought map itself. Map production involves collaboration between senior-level technical specialists from institutions in all nine northeastern states and several federal entities. The process for establishing the Monitor has facilitated important technical and institutional enhancements for the development of policies and plans aimed at drought risk reduction, as will be described in further detail in Chapters 4 and 11.

The second instrument, or set of instruments, consists of developing preparedness interventions with well-defined and negotiated actions to be implemented when a drought event begins and as it evolves through time and space. Drought preparedness plans (which include elements of risk mitigation and contingency planning) define the types of actions to be taken for the different stages or intensities of a drought (i.e., from beforehand to the first signals of drought, and on to extreme and exceptional droughts), as well as the initiation and termination conditions for each stage. The plans characterize drought impacts and vulnerabilities, key institutional actors, and planning measures for mitigating drought risk, as well as emergency responses. Their contribution toward better articulation among the different sectors and levels of government involved is illustrated in Chapters 5 and 6.

3.5 Conclusion

The initial steps taken by Brazil over the past few years, with the assistance of national and international partners, to usher in technical upgrades (drought monitoring and preparedness plans) have been both supported by and contributed to institutional ones that were equally, if not more, important than the technical improvements themselves. Chapters 4 through 6 will illustrate the dialogue and process associated with the Drought Monitor and Preparedness Plans that have helped to structure and facilitate a more permanent institutional approach and response to this recurrent phenomenon in the Northeast region as well as to improve DRM integration within and among federal and state institutions. However, despite the progress these upgrades represent, as Chapter 8 will show, these improvements represent just the initial stages of the paradigm shift that still needs to advance in the years ahead.

4

The Technical and Institutional Case: The Northeast Drought Monitor as the Anchor and Facilitator of Collaboration

Eduardo Sávio P.R. Martins, Carmen Molejón Quintana, Maria Assunção F. Silva Dias, Robson Franklin Vieira Silva, Bruno Biazeto, Gisela Damm Forattini, and Julia Cadaval Martins

CONTENTS

4.1 Limitations of the Traditional Approach to Drought Monitoring........37
4.2 The Possible Solution Based on the Drought Monitor: A Change in the Paradigm........39
4.3 Drought Preparedness Plans and the Monitor........45
4.4 The Challenges and Use of the Monitor........47

4.1 Limitations of the Traditional Approach to Drought Monitoring

As observed in Chapter 1, the semiarid Northeast region of Brazil occupies a very large area and one of its principal characteristics is deficiency in rainfall. This is a historical problem. Even though the first recorded drought in this region was in 1583 (Campos 2014), these events only began to be systematically recorded in the nineteenth century. The first occurrences were noted in connection with their associated impacts, while more recent ones are based on meteorological observations.

To illustrate the climatic variability typical of the region, Figure 4.1 shows the intra- and inter-year distribution of rainfall between 2008 and 2015 for the northeastern state of Ceará, together with the mean annual precipitation (monthly climatology). Years in which the rainy season was below and above the average are presented in red and blue, respectively. This pattern of wet and dry years, even if not precisely the same, was observed for the Northeast semiarid region as a whole. In essence, responses to this variability have been reactive, both in years of above-average rainfall (e.g., increased

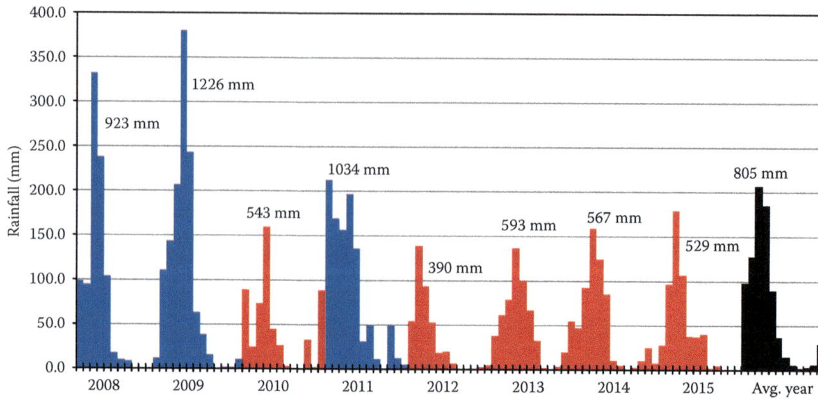

FIGURE 4.1
Intra- and interyear distribution of rainfall for the state of Ceará for the 2008–2015 period. Months are depicted for each year from January to December (left to right). Years in which the rainy season was below and above the average are presented in red and blue, respectively. The annual average amount and distribution are presented in black at the right of the figure.

livestock herds, increased planted area) and in those of comparative water scarcity (drilling of wells, emergency water supply systems, etc.)

The heavy investments made in water-related infrastructure in the Northeast region in the 1990s and 2000s appear to have been a response to the impacts of the recurrent droughts, at least with respect to those specifically involving water systems (responding to urban demand, expansion of irrigated agriculture in perimeters, etc.). However, the present (i.e., 2010–2015) extended drought, which is the most severe of the past 80 years, has demonstrated that decades of structural solutions, while necessary, are insufficient to overcome multiple successive years of below-average rainfall. This fact revealed a need for a deeper reflection in terms of public policy in Brazil, as adequately addressing regional fragility in the face of droughts is not restricted to infrastructure solutions, but requires short-, medium-, and long-term management of droughts and their effects as well.

As has been stressed in Chapters 1 through 3, the focus of drought management in Brazil has historically been reactive in nature, with response mechanisms defined during drought periods by temporary committees at the federal level. Because of their institutional weaknesses in terms of both human and financial resources, the states are generally limited to implementing programs designed by the federal government. Furthermore, the federal and state institutions responsible for meteorological, hydrological, and agricultural monitoring and forecasting have generally acted independently, generating a large number of different products in relation to drought monitoring. As a result, state and federal agencies often disagree on the severity of a drought in a specific locality, leading to different perceptions regarding the need for the mobilization or demobilization of resources, particularly emergency ones. Better coordination among the federal and state entities involved in drought management is clearly needed.

Overlapping of responsibilities occurs not only across different levels of government, but it is possible to identify single institutions that generate more than one product with the same purpose and monitoring nearly the same drought dimension, for example, the meteorological aspect. One of the reasons for this is that these products are the result of individual sector-specific initiatives and do not emerge from a national drought policy. Another characteristic of these products is the lack of integration of data generated by the multiple federal and state monitoring networks in the country.

The diversity of institutions and outputs traditionally involved in drought monitoring in the Northeast region has tended to impede improvements in the timeliness and level of warnings with respect to the severity of drought conditions in the region. This observation goes somewhat against the findings of several studies in the area of governance that associate advantages to the creation of redundancies among institutions (Landau 1969; Scott 1985) and reinforces the argument that redundant elements can be used in many different ways in diverse systems (Felsenthal 1980; Lerner 1987). In the present instance, however, useful information is generated but not utilized as effectively as it could be for drought management in the region. This leads to the conclusion that what is needed is not solely a new and improved product, but also a process that effectively coordinates the various federal and state agencies involved in monitoring and forecasting droughts and responding to their impacts.

A better understanding of the fragilities of the Northeast semiarid region is also required prior to the planning and implementation of water infrastructure investments. This passes through definition of a sustainable development model that is appropriate for the region and incorporates climate information into planning activities, both for large infrastructure, taking into account likely climate variability in the areas where it will be located over the medium and long run, and for its management in the short term.

4.2 The Possible Solution Based on the Drought Monitor: A Change in the Paradigm

As noted in the preceding chapters, this situation has recently led to a dialogue in Brazil concerning how to improve drought policy and management. The need for a better coordinated government response to the occurrence of droughts, involving all three administrative levels—federal, state, and municipal—and not only in a short-term reactive form, but also a longer-term proactive one, has led the Ministry of National Integration (MNI) to encourage elaboration of a more structured National Drought Policy.

In the past, discussions regarding the need for a national drought policy gained or lost momentum in accordance with the stages of the drought cycle, with only limited incremental progress being made toward more proactive drought management. Initially, given the complexity of the topic, the government decided to focus on the first and most fundamental pillar of drought preparedness: monitoring. As indicated in Chapter 3, improved monitoring is the key to facilitating a change of paradigm. However, it is only one element in this regard.

As a first step, a more adequate drought monitoring model was sought, one that presented enhancements in relation to the conventional monitoring hitherto undertaken by various institutions at both the national and state levels. The improved monitoring model selected was inspired by drought management activities in Mexico and the United States that bring together information from both federal and state sources to produce a single map of regional drought conditions, as will be explained further in Chapter 7. As implementation of this model requires stronger articulation among state and federal institutions, it was decided to limit its execution in the first instance to the Northeast region. Just as this process and the associated institutional cooperation that supports it, the resulting map (i.e., the final monitoring product) seeks to promote a common definition and understanding of the drought, as well as to increase the efficiency and effectiveness of public policy responses. This new model is referred to from here on as *the Northeast Drought Monitor*, the *Drought Monitor*, or simply *the Monitor*.

The fundamental difference between the Drought Monitor and conventional systems is the emphasis it gives to the participatory process that leads to the creation of its key product (i.e., the monthly drought map), as opposed to an automatic generation on the basis of numerical calculations and drought indicators, which do not necessarily reflect the intensity and/or nature of the drought as experienced in any specific locality. Indeed, the majority of traditional drought monitoring products focus merely on the meteorological dimensions of droughts. However, below-average precipitation in a given period, for example, can nonetheless result in above-average harvests and above-average discharges into reservoirs, depending on its temporal distribution. Thus, to inform public policy decisions, monitoring droughts only from a meteorological perspective is not sufficient.

Accordingly, the Northeast Drought Monitor entails a process that counts on the participation and collaboration of a number of climate-related institutions, including in both the water resource and agriculture sectors of the states, as well as of selected federal institutions, with the aim of identifying the degree of drought severity in various parts of the region in its meteorological, hydrological, and agricultural dimensions. As part of the process of designing the Drought Monitor, an institutional diagnosis of the pertinent federal and state agencies was carried out, and they were invited to participate in regional workshops to discuss its operational and institutional design as well as to accompany progress during its *experimental* implementation phase, which was carried out between August 2014 and August 2015. The Northeast Drought Monitor was subsequently launched as *operational* at the beginning of 2016 (http://monitordesecas.ana.gov.br/).

The road toward operational implementation of the Monitor, however, was not trivial. Among the reasons for this were that the majority of the institutions consulted initially felt that their current drought monitoring procedures were already adequate. To encourage different institutions to participate in the Monitor, several regional workshops were held in the Northeast region and complemented by bilateral meetings with the core team of the then-denominated *Monitor Project*. Through these means, the team aimed to collect each institution's perspectives on the possible value added of a common Drought Monitor, and incorporate those into its development process.

Despite the initial challenges and difficulties, this participatory process led to integration of all the relevant regional databases, thereby permitting calculation of different drought indicators, and bringing together auxiliary products from other sources, such as remote sensing. Combining the indicators and auxiliary products is the responsibility of a lead Northeast *author* institution* and results in the generation of a preliminary drought map. In order to identify discrepancies between this map and local evidence, the draft is then submitted to a local validation process involving stakeholders that are directly experiencing drought impacts. This is one of the most important steps in the process because the draft map still entails some inaccuracies, principally because of the low density of the regional monitoring network, which is insufficient to reflect drought intensity in every locality due to the spatial variability of rainfall, soils, vegetation, and land uses, among other characteristics. This validation exercise is repeated until a final map based on consensus among all the institutions involved (i.e., both the initial *author* and the *validators*) is produced. Due to the collaborative nature of the process among institutions at both the state and federal levels, a common awareness is consolidated with respect to the severity of the drought impacting various parts of the region.

For the reasons highlighted above, the Drought Monitor aims to have the following characteristics:

- It is the result of a set of concerted monitoring actions between the federal and state institutions.
- Because of its multi-institutional framework and of its complex operational aspects, the process should start as simply as possible and evolve over time (by the inclusion of new indicators, monitoring

* During the experimental phase of the Monitor it was determined that for each month one state institution (known as the *author*), in rotation, would lead the process of drawing the Monitor Map.

stations, auxiliary information derived from remote sensing and modeling, etc.) as experience is gained.

- It should include coauthorship of the states elaborating the final product (i.e., the monthly drought map), together with the supervision of a federal entity.
- It should include local validation of the drought conditions identified in the process of building the Map; thus, the Drought Monitor should make use of a network of local validators responsible for confirming or contesting the projected severity of the drought indicated by calculating the various drought indicators (i.e., meteorological, hydrological, and agricultural ones).
- It should facilitate through its categories the definition of triggers, or threshold values of the drought indicators, which lead to concrete government responses to the drought in accordance with its observed severity at different locations in the region.

The collaborative nature of the process among the institutions at the federal and the state levels, together with local validation, strongly increases the likelihood of consensus between them with respect to the characterization of the severity of the drought and its evolution over space and time, thus minimizing divergent conclusions regarding the need to mobilize or demobilize emergency resources for a specific area in the region. Table 4.1 presents the drought categories and their associated impacts utilized by the Drought Monitor in order to improve the definition and characterization of the droughts in Northeast Brazil. The regional drought maps were developed

TABLE 4.1

Examples of Drought Categories and Associated Impacts That Are Tracked Using the Drought Monitor

Category	Percentile	Description	Possible Impacts
D0	30	Abnormally dry	Going into drought: short-term dryness slowing planting and growth of crops or pastures. Coming out of drought: some lingering water deficits; pastures or crops not fully recovered
D1	20	Moderate drought	Some damage to crops, pastures; streams, reservoirs, or wells low; some water shortages developing or imminent; voluntary water-use restrictions requested
D2	10	Severe drought	Crop or pasture losses likely; water shortages common; water restrictions imposed
D3	5	Extreme drought	Major crop/pasture losses; widespread water shortages or restrictions
D4	2	Exceptional drought	Exceptional and widespread crop/pasture losses; shortage of water in reservoirs, streams, and wells creating water emergencies

in an integrated and systematic manner with a view toward both drought monitoring and local impact validation. The Drought Monitor is also a key tool to support the dialogue between the states and the federal government about addressing drought risks and conditions in the semiarid region as well as for drought preparedness planning, thereby creating a platform for consensus building and institutional integration.

The Northeast Drought Monitor, thus, constitutes a graduated decision support instrument, aiming to enhance both drought preparedness and response to its effects, based on evolving indications of drought severity and likely duration, observing the characteristics listed above. This process, moreover, supports both drought policy formulation and proactive drought management on the ground. In its most visible form, the Monitor, at this initial stage, seeks to produce a monthly map that describes the current drought stage throughout the region in terms of the categories identified in Table 4.1. This process furnishes a more subtle and objective definition of a drought on the basis of a protocol that weighs different indicators in a composite index for the whole Northeast region, with the drought severity level depicted being subject to local validation. Figure 4.2 shows an example of the Map produced during the experimental phase of the Monitor for July 2015, following the categories described in Table 4.1.

The coordination and sharing of efforts to produce the Drought Monitor has generated positive externalities, as this initiative has also resulted in greater integration of the various federal and state institutions involved.

The institutional cooperation that supports the entire process was structured in two levels: the strategic and the operational. The strategic level is coordinated by a Steering Committee (*Conselho Diretor*) that reports to the federal government, a role presently exercised by the MNI and which could eventually be exercised by the Chief of Staff of the Presidency of the Republic (*Casa Civil*). The steering committee includes among its members representatives of the Ministries of National Integration; Environment; Agriculture, Fisheries, and Food Supply; Agrarian Development; and Science and Technology. It also has representatives from key agencies linked to these ministries, as well as from the nine states of Northeast Brazil that participate in the Monitor. The steering committee may also eventually be leveraged to strengthen the other two pillars (Wilhite et al. 2005) of the National Drought Policy.

The operational level, in turn, is led by a central institution, a role initially performed by the Ceará Foundation for Meteorology and Water Resources (FUNCEME) during the experimental phase. This institution worked in close coordination with an advisory board (*Corpo Consultivo*) composed of representatives of each state and federal institution involved in drought management in the Northeast region, which makes recommendations to be implemented at the regional level in the context of the Monitor.

The central institution assumes the role of executive secretariat of the Monitor, acting in accordance with the decisions of the steering committee

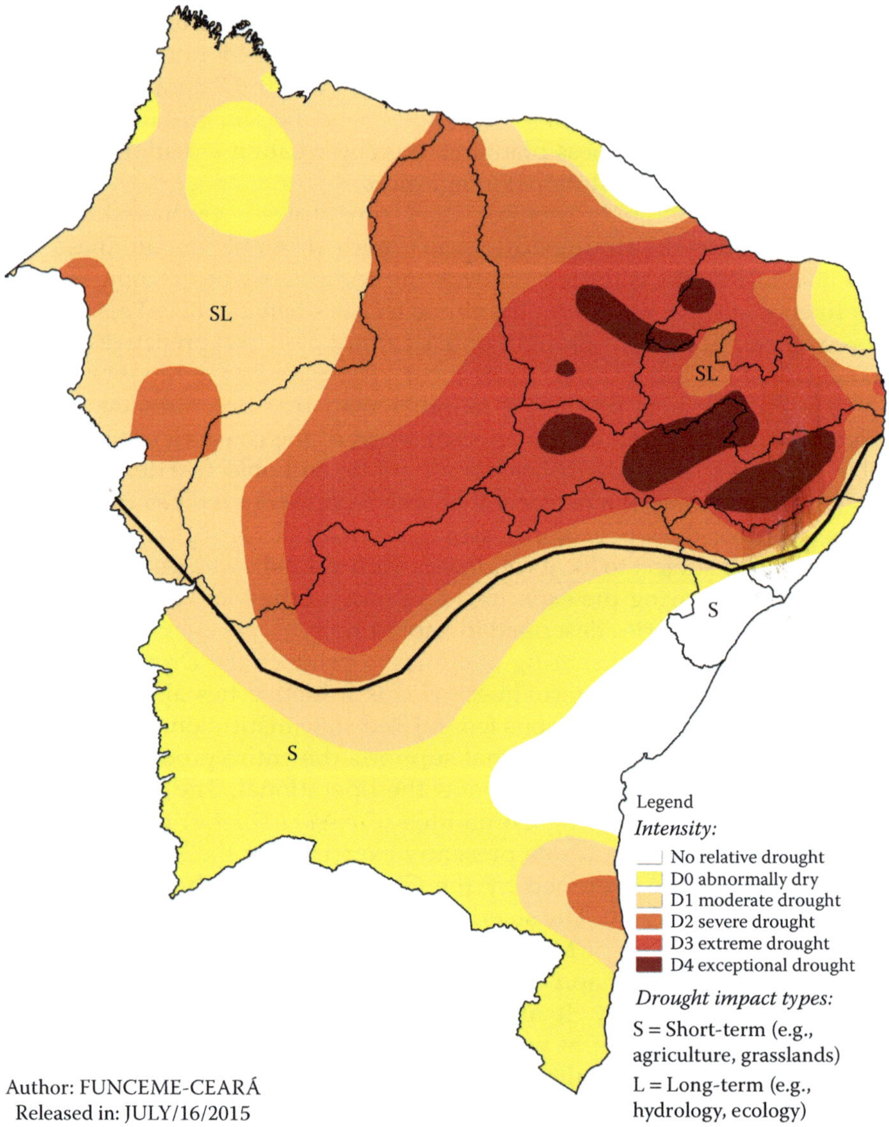

FIGURE 4.2
Experimental map of the Drought Monitor for Northeast Brazil for July 2015. The drought categories are calibrated for severity, similar to those that appear in Table 4.1. The areas that are indicated as in short-term drought are marked as "S," while those indicated as in long-term drought are identified with "L." Some areas appear as both short- and long-term drought and are highlighted with an "SL" designation. Impact Duration: S = Short-term (e.g., agriculture, pastures); L = Long-term (e.g., hydrological, ecological). (From: Northeast Drought Monitor/Author: FUNCEME.)

and working in harmony with the advisory board. As a result of discussions among the institutions involved, and as an independent regulatory agency whose mandate includes the possibility of taking on such responsibilities, the National Water Agency (ANA) has now assumed the role of the central institution/executive secretariat on a more permanent basis, guaranteeing institutional stability for the Monitor despite inevitable changes over time in federal administrations. In addition to the central institution, *ad hoc* commissions are planned to help oversee specific themes of interest for the Monitor.

Drought risk planning and associated response and mitigation measures—elements that are connected with the third pillar of drought preparedness and addressing drought risks in the semiarid region (Wilhite et al. 2005)—are not incorporated into the Drought Monitor *per se*, but will ultimately have a strong interaction with it. Various strategies are being considered to make this interaction feasible, for example, through an institutional arrangement that permits representatives from state-level drought committees to oversee the Monitor's advisory board. These committees are strongly involved in the planning and implementation of drought response measures at the individual state level. Such planning still occurs largely in response to demands from the specific localities impacted by the drought, but the committees' participatory interaction with the Monitor is expected to contribute to a change in their response to the problem, from a crisis management to a risk management one. Furthermore, including these committees in the institutional arrangements for the Monitor may favor their establishment on a permanent basis.

4.3 Drought Preparedness Plans and the Monitor

The Drought Monitor reflects the physical or natural drought, while the shortage of water caused by humanity should be indicated by elements complementary to the Map. In an effort to establish the connections between the Monitor in its physical dimensions (i.e., meteorological, hydrological, and agricultural) and drought preparedness plans for the systems to be managed, a pilot program for the region was launched with several case studies. The idea behind this initiative was to demonstrate the use of concrete strategies and tools for proactive drought management through the conception and development of both the Drought Monitor and drought preparedness plans. The relation between the Monitor and the preparedness plans depends significantly on that between the physical and natural drought,* portrayed by the Monitor, and the operational one,†,‡ indicated by the preparedness plan.

The pilot program for the Northeast region involved elaboration of five preparedness plans for case studies in four different contexts or sectors: urban water supply; river basin planning; management of small multiuse reservoirs; and rain-fed smallholder agriculture. Together with the Monitor, these plans sought to make drought preparedness tools accessible to decision makers and to demonstrate the value of the paradigm change in terms of proactive drought management. Accordingly, the case study teams and their partners sought to consolidate each plan, in both its elaboration and implementation, following the *three pillars* mentioned in Chapter 3. To the extent possible, they also sought to connect the Monitor and its drought categories (i.e., D0–D4) and the specific policy and management actions proposed, as triggered by these categories, in the respective drought preparedness plans.

These plans are starting to be implemented at the local level with the intention that they be used to guide short-term decisions as the drought develops and to help orient longer-term investments, while also seeking to provide solutions for underlying vulnerabilities and to mitigate risks of future droughts. The next step will be for these initial drought preparedness plans to inform discussions between the federal and state governments about how to expand such approaches throughout the Northeast region. Currently, ANA is cooperating with FUNCEME to expand these plans to other parts of the region. Examples of these plans are detailed further in Chapters 5, 6, and 12.

The plans characterize drought impacts and vulnerabilities, the principal institutional actors involved, planning measures for the mitigation of drought risks, and emergency responses. Some of the plans were not capable of defining policy and management actions while the drought was progressing toward more severe levels (i.e., D3 and D4), as in the case of smallholder rain-fed agriculture, while others, such as the two urban water supply cases, formulated a range of actions to be triggered once a water system progresses to the next level of drought severity. None of the plans were able to extract information directly from the Monitor to inform policy actions or triggers because, in its initial phase, it did not yet possess the breadth and diversity

* As stated at the beginning of Chapter 1, it is important to distinguish drought from water scarcity. Drought is a natural phenomenon, but water scarcity occurs when the shortage of water is caused by humankind using more water than naturally available. The Monitor represents the drought with natural causes, while intending to include also some information on the human influence on water availability.

† The reservoirs that respond to urban demands and irrigated agriculture will be represented by circles divided in half, with the left-hand portion presenting the drought severity indicator with respect to urban water supply, while the right-hand portion refers to the drought severity indicator regarding irrigation schemes.

‡ Operational drought could be thought of as a drought resulting from a sequence of decisions on how to manage a system, such as a reservoir system. An operational drought can speed up the impacts of a physical drought over a system. The system here is referred to as a managed system, since it is subject to human-made decisions.

of indicators to justify a direct link with them. However, most of these plans adhered to the new D0–D4 classification of drought severity and intend to use this system of categorization to provide feedback to the Monitor in order to inform its characterization of the drought (e.g., the levels of reservoirs in the urban case associated with D0–D4 can help the Monitor define the severity of the drought in these areas). All of the plans nonetheless highlight the needs for continuous iteration and to reinforce their ties with the Monitor in the future.

4.4 The Challenges and Use of the Monitor

The Northeast Drought Monitor, as a process, organizes and provides more consistent and locally validated drought-related information, seeking with this to achieve a reduction in political pressure and interference as well as to avoid subjective judgments, for example, at the time of declaring a drought situation in a particular municipality. And, as the process is a collaborative one among multiple federal and state institutions, it is expected that greater consensus regarding the status of an evolving drought is more likely to occur. As a result, those involved with this process have renewed hope that there will be greater agility and consistency in the definition of government responses to droughts through the triggers defined for each of the water systems analyzed. However, effectiveness of the changes that can be introduced by the Drought Monitor depends on overcoming an important obstacle: the information made available through the Monitor will not be translated into decisions unless clear preparedness plans with previously defined actions exist to guide decision making.

The institutional collaboration that supports the Drought Monitor is particularly concerned with its effective use for drought management at both the strategic and operational levels. The Monitor's high-level strategic coordination has the intended purpose of defining general targets and evaluating performance with respect to the three pillars of drought preparedness reflected in a national drought policy. The evaluation process will, thus, include an assessment as to whether progress is being made in the use of the Drought Monitor to trigger response actions and if the Monitor as a whole is evolving in the desired way. Operational coordination of the Drought Monitor, in turn, will not only guarantee the maintenance and improvement of routine activities and of the processes involved, but promote expansion of its use by various sectors and the society in general. This also includes the dissemination, training, and education activities associated with the Monitor.

The momentum experienced during the Monitor's experimental phase, and now in the early stages of its operational phase, needs to be capitalized upon in order to promote its use and application throughout the Northeast region.

This is not an easy process, but considering the differences between the Monitor and more traditional approaches and their respective outputs, there is hope that this process will be internalized over time into proactive drought management efforts at the federal, state, and municipal levels.

Chapter 11 of this book explains in more detail the different steps involved in the collaborative process used by the Drought Monitor to produce the monthly drought maps for Northeast Brazil.

5

Water Supply and Management Drought Preparedness Plans

Francisco de Assis de Souza Filho, Marcos Thadeu Abicalil, Paula Pedreira de Freitas de Oliveira, and Cybelle Frazão Costa Braga

CONTENTS

5.1 Conceptual Framework...51
5.2 Drought Planning on Multiple Spatial, Temporal, and Sectoral Scales..53
5.3 Drought Preparedness Plan for a River Basin.......................................55
5.4 Drought Preparedness Plan for a Hydrosystem 59
5.5 Drought Preparedness Plans for Urban Water Utilities........................61
5.6 Final Observations...64

Droughts are an intrinsic characteristic of climate and occur in many countries, regions, and places. The temporal patterns of climate variation, the level of aridness, and the social and environmental uses of water contribute to determine their impacts. In addition, as observed in Chapters 1 through 4, until recently, government planning and interventions in relation to drought events have generally occurred in a reactive fashion, with drought-related actions usually defined after these events are already well-advanced.

Over the years, a range of measures have been implemented, from the household to the national levels, to minimize drought effects. These actions have been applied in emergency situations once an event has begun, have not been continuous, and are often forgotten as soon as the rain returns. The water storage capacity of the state of Ceará is illustrative of this fact. Figure 5.1 shows the increase in this capacity following the 1919, 1932, 1958, and 1979–1983 droughts, demonstrating the *hydroillogical* cycle referred to in Chapter 1 (Wilhite et al. 2005). More recent social protection policies seeking to boost food security (e.g., through *Bolsa Família*) and sectoral policies such as those for water resources (i.e., demand management and the construction and operation of water storage and distribution facilities) have reduced social vulnerability to the drought. However, severe water supply impacts on urban centers, rural populations, and both irrigated and rain-fed agriculture have continued to occur, as the present drought in the Northeast region clearly shows.

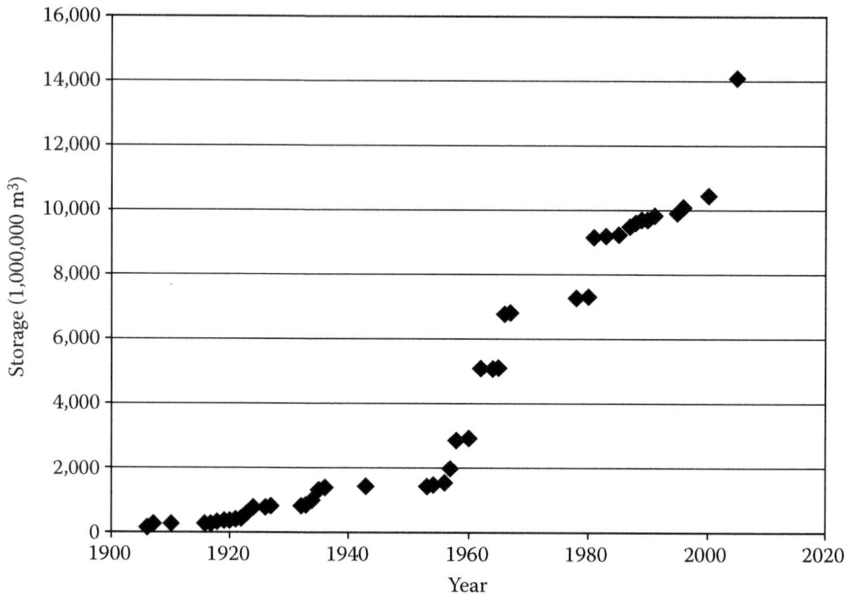

FIGURE 5.1
Evolution of the capacity of water storage in reservoirs in Ceará.

Drought impacts affect different sectors and require measures in different policy areas (water supply, sanitation, agriculture, industry, fisheries, energy, transportation, etc.). Plans that respond to the multiple spatial and sectoral dimensions of drought impacts need to be elaborated.

The risks associated with droughts are the product of local exposure (i.e., the probability of occurrence) and vulnerability of the locality or localities affected. For this reason, the development of a drought preparedness plan is a significant step in the adoption of a proactive management process. This plan should have as its pillars continuous monitoring and early warning systems, risk assessment, and the definition of actions for drought preparedness, mitigation, and response. The continuous monitoring is associated with the use of appropriate indicators and indices that are linked to specific impacts, triggers, and development of a decision support system (DSS).

Planning properly and implementing them during nondrought periods can improve the government capacity to anticipate water scarcity in an effective way. Thus, planning can reduce and, in some cases, avoid impacts, minimizing physical and emotional suffering in the process. Drought preparedness planning is a dynamic process that should incorporate both traditional and emerging techniques and take socioeconomic, agricultural, technological, and political trends into account (Wilhite 1996). Planning for droughts should not begin or end with the publication of a drought-related plan, but it should establish political–institutional momentum for the maintenance and upscaling of the drought management process.

Consequently, an adaptive and evolutionary planning process is proposed for drought risk management. This process is *adaptive* because planning (1) needs to recognize the social capital (i.e., social networks together with shared norms, values, and understanding that facilitate cooperation among and within groups) that provides its context and is important for the design of the solution; (2) should be sufficiently flexible to deal with the uncertainties inherent in complex systems, such as socionatural ones; and (3) is strategic for the provision of greater resilience to droughts. It needs to be *evolutionary* because the process requires continued refinement of (1) the methods involved and (2) the quality of the measures foreseen, starting from solutions having low social and technical costs and advancing toward methodological alternatives with greater technical and organizational requirements. This social, technical, and organizational learning is inherent in the participatory planning process, and its appropriation by an organization modifies it, requiring the updating of planning activities for the new organizational and technical context. The development of drought preparedness plans described in this chapter is based on this approach.

This chapter presents a general overview with respect to preparedness planning for droughts in Brazil. Further, this chapter summarizes a conceptual framework; describes the levels and types of drought-related planning; and presents three specific examples, one for a river basin, another for a *hydrosystem*, and the third for an urban water utility, followed by brief final observations.

5.1 Conceptual Framework

Historically, actions in relation to droughts have occurred exclusively as reactions to such events as characterized in the *hydroillogical* cycle mentioned in previous chapters. It is *illogical* because, over time, droughts have been understood as random events that require *emergency actions* that are only considered after the onset of a drought, resulting in reactive rather than proactive measures to address it. This process has led to the elevation of costs and drought impacts to the extent that they favor only temporary relief for the affected populations and reinforce their dependency on local political elites and the government, as pointed out in previous chapters.

Planning as an instrument that involves prior and systematic deliberations and actions is the appropriate tool to break the *hydroillogical* cycle. Otherwise, droughts only enter decision makers' agendas after they become sufficiently serious and exit these agendas once they end. This tendency results in significant inefficiencies in drought-related actions. Thus, it is recommended that planning be integrated within a broader drought risk management approach. In short, a continuous planning process is required instead of the mere preparation of a plan. Learning from the experience of previous droughts and developing efficient and effective organizations and strategies to manage them should compose the drought management cycle.

Risk management requires explicit recognition of tolerable risk levels (TRLs) for the systems and sectors involved. Drought management measures are designed to meet these risk levels, or lower ones. For situations in which the present system is unable to offer patterns compatible with the TRL, mitigation and adaptation actions that lead to the desired levels of the system in the form of long-term strategic plans (e.g., a water security plan) should be foreseen. There is a trade-off between the TRL, the costs associated with these risk levels, and the willingness of society to pay for their reduction. Operational plans for drought management, in turn, seek to define measures that can be implemented within the present configurations of the systems under consideration. The plans discussed below are operational in nature.

The risks associated with droughts are distributed unequally across social groups. The poorest are the most vulnerable and, consequently, are subject to the greatest risks. Therefore, in defining the measures to be implemented, in addition to economic efficiency, drought-related plans need to consider social equity concerns. The specific plans discussed in this chapter are for water resources for multiple purposes (human consumption, irrigation, etc.). Operational plans for drought management involving these aspects need to be considered on three spatial levels—for the entire river basin, for specific hydrosystems, and for water supply to urban water utilities (Figure 5.2). The hydrosystem (composed of both surface and subterranean water supply and water demand sources) can be associated with more than one river basin in the case of interbasin water transfers.

Drought management in its broadest sense should integrate (1) the general planning of water systems with actions to guarantee the balance between water availability and future demands; (2) operational rules of water systems under normal conditions and rules for drought scenarios; and (3) management strategies and operational scenarios to mitigate impacts under drought conditions (González and Morcillo 2007). Droughts are associated with system failures due to insufficiency of water resources to meet demands because of prolonged episodes of low rainfall or reduced flows. These are often the result of an imbalance between the evolution of demand and existing supply due to planning or operational shortcomings in terms of system infrastructure. However, it is frequently difficult to quantify the relevance of each of these causes in a given drought event. Because of this difficulty, some authors (Andreu and Solera 2005; González and Morcillo 2007) use the concept of *operational drought* to define the event independently of its primary cause. Once the need for different uses of water have been identified, including by the ecosystem, if the resulting evaluation of the natural sources and management and the operational system do not meet these requirements, this can be referred to as an operational drought.

As alluded to earlier, the development of a drought preparedness plan should consider the three basic pillars described in Chapter 3: (1) monitoring,

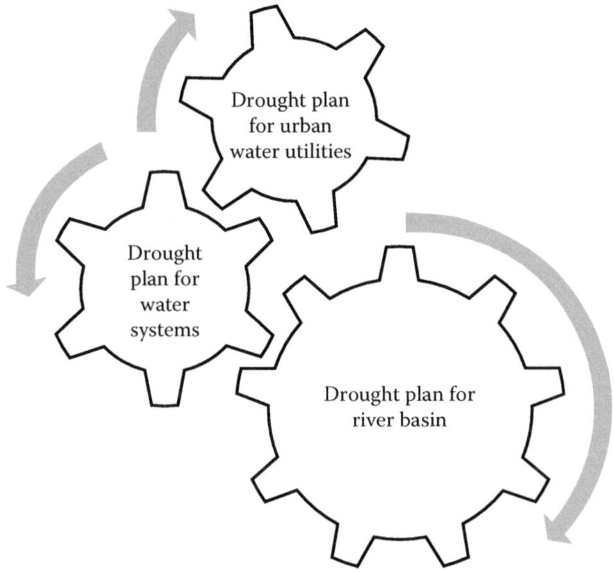

FIGURE 5.2
Drought planning at multiple levels, and the notations inside the gears (from top to bottom) for the *drought plan for urban water utilities, drought plan for water systems,* and *drought plan for river basin* described in this chapter.

forecasting, and early warning; (2) risk and impact assessment; and (3) preparedness, mitigation, and response actions.

Monitoring in this instance is associated with the use of appropriate indices or indicators that identify the degree of drought severity. The degrees of severity, in turn, serve as triggers to initiate risk management actions (i.e., preparedness, mitigation, and response). Development of a DSS for this purpose is desirable. In the risk and impact assessments, potential effects on the locality considered should be identified, inventoried, and monitored in order to determine what is at risk and why. Drought response actions should be proactively implemented to reduce the risks and increase social capacity to address them.

5.2 Drought Planning on Multiple Spatial, Temporal, and Sectoral Scales

The occurrence of meteorological droughts is associated with temporal and spatial variations in precipitation. This variation can result either in generalized or widespread droughts when they occur in large areas or more localized ones. Human societies seek to adapt to these patterns in order to cope with the variations involved. Governments in the Northeast region often explicitly express their interest in strengthening societal capacity to adapt to the climate in the form of *coexisting with drought* or *plans to combat drought*. However, these approaches usually tend to have a short-term vision in order to deal with the current drought, again, as noted in Chapters 1 and 2.

Human populations and economic sectors possess different levels of vulnerability to water-related stresses, leading to potentially different impacts on them for the same natural drought event. Drought-related planning should, thus, also seek some degree of cross-sectoral coordination through a national drought policy. Both past planning experiences of a reactive character and more recent ones following a risk management approach should be assessed with a view toward configuring a more comprehensive drought preparedness planning system. Some of the potential levels of planning in this system and their purposes (see also Figure 5.3) are as follows:

- *State drought plan*: Establish strategic planning at the state level with a view toward the definition and coordination of drought actions and resources.
- *Drought preparedness plan for a river basin*: Provide a river basin with a planning instrument with guidelines, strategies, actions, and information for mitigation, preparedness, and responses to drought events, with a special focus on the definition of water use rules in years of scarcity, as well as water supply strategies for the rural and urban populations and for economic and environmental uses.

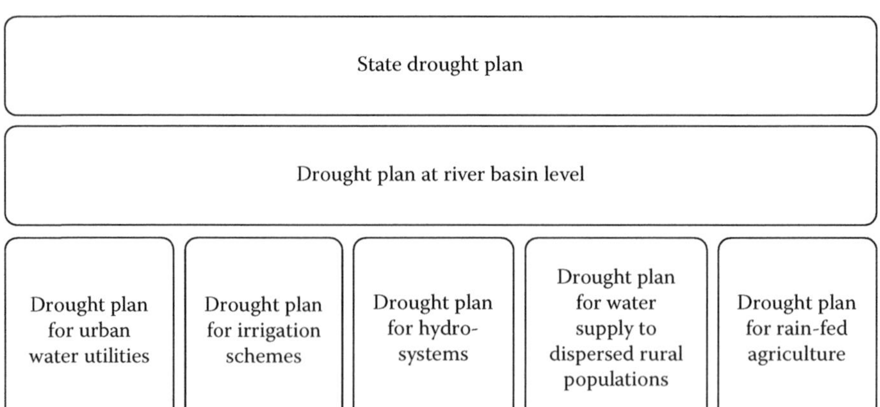

FIGURE 5.3
Scales of drought planning.

- *Drought preparedness plan for a water system*: Define allocation rules associated with system operation and drought levels through procedures involving triggers.
- *Drought preparedness plan for urban water utility*: Define operational measures to be undertaken by the respective water utility(ies) to mitigate drought risks and distribute water to the urban area(s) in question.
- *Drought preparedness plan for irrigation schemes*: Define operational measures to be undertaken by irrigation scheme managers to mitigate drought risks and distribute water within the irrigation scheme.
- *Drought preparedness plan for water supply to dispersed rural populations*: Define strategies and operational mechanisms for the provision of water to dispersed rural populations.
- *Drought preparedness plan for rain-fed agriculture*: Define agricultural strategies and management practices to reduce drought impacts on rain-fed agriculture.

These levels of planning should be viewed as instruments of a national or subnational drought policy, based on a strategic vision together with planning and management activities that reduce the vulnerabilities and risks associated with droughts. In addition, these instruments should be harmonized with plans for other sectors such as water resources, sanitation, irrigation, urbanization, and climate change adaptation, among others.

5.3 Drought Preparedness Plan for a River Basin

In Brazil, the river basin is defined as a planning unit through national legislation (Law 9433/1997). Thus, it is essential to adopt this scale for drought management to ensure that drought preparedness plans are harmonized with basin plans.

The Piranhas-Açu river basin, with a drainage area of about 43,681 km^2, and shared by Paraíba and Rio Grande do Norte states, was selected as the case study at the river basin level, and, in this context, a protocol* for drought preparedness in the basin was developed on the basis of the concepts introduced above.

The objective of this protocol for drought preparedness was to provide the Piranhas-Açu river basin with guidelines, strategies, and information for the mitigation, preparedness, and response to drought events, including in more detail the following:

* For the Piranhas-Açu river basin, it was decided to call this instrument the *protocol* for drought preparedness in order to avoid a conflict with the Basin Plan. This protocol was subsequently incorporated into the Basin Plan.

- Drought monitoring
    - Collection of hydrometeorological data
    - Calculation of drought indices/indicators in accordance with the basin's characteristics
    - Establishment of drought monitoring indicators
    - Establishment of drought categories
    - Monitoring of hydrometeorological variables
    - Development of a DSS
- Assessment of impacts and vulnerability
    - Identification of the main impacts on the basin
    - Vulnerability assessments
    - Monitoring/inventory of impacts and vulnerability
- Preparedness, mitigation, and response actions
    - Proactive programs and measures to reduce risks and increase capacity to confront the drought
    - Proposed actions in relation to the impacts
    - Water resource supply and demand management actions

The combination of a given drought category and the impact and vulnerability assessments indicates the urgency of preparedness and mitigation for different drought scenarios, and consequently, the need to implement strategic, tactical, or emergency actions (Figure 5.4).

The development of a drought preparedness plan should be participatory, involving all of the basin's institutional stakeholders (Wilhite et al. 2005). Accordingly, elaboration of the protocol for drought preparedness entailed establishment of strong articulation with the federal and state water resource management institutions for the basin, including the National Water Agency (ANA), the National Department for Works against Droughts (DNOCS), the Environment and Water Resource Secretariat of Rio Grande do Norte State (SEMARH-RN), the Water Management Institute of Rio Grande do Norte State (IGARN), the Executive Water Management Agency of Paraíba State (AESA), and the River Basin Committee of the Piancó and Piranhas-Açu River Basins (CGH-PPA). All phases of the protocol's elaboration were discussed and validated with all of the institutional actors involved.

FIGURE 5.4
Drought preparedness planning process for the Piranhas-Açu river basin.

TABLE 5.1

Drought Scenarios

Stage	Drought Trigger	Response Targets
Warning	Below Target Level 1	10% Reduction in Consumption
Moderate drought	Below Target Level 2	20% Reduction in Consumption
Severe drought	Below Target Level 3	30% Reduction in Consumption
Extreme drought	Below Target Level 4	60% Reduction in Consumption

The basin diagnostic in relation to the drought was established on the basis of the monitoring of hydrometeorological variables (i.e., precipitation, reservoir levels and discharges). The drought categories were determined on the basis of several indicators, the standard precipitation index (SPI), the standard discharge index (SDI), and the status index (SI), that is, the level of water storage in the reservoir. These indicators were considered triggers for the definition of the drought categories and hence their severity (Table 5.1).

The vulnerability assessment can orient decision makers for the adoption of drought preparedness measures since it indicates the areas that require greatest support. For the Piranhas-Açu basin, stakeholders decided to calculate vulnerability as a function of exposure, adaptive capacity, and sensitivity based on indicators. For the selection of preparedness, mitigation, and response actions related to the drought categories, the following *thematic axes* were considered:

- Institutional
- Communication
- Legal and normative
- Water allocation and water use permits, including hydroeconomic negotiated allocation, that is, considering economic values of water allocation and trade-offs among uses
- Monitoring
- Water systems operation
- Integration with the São Francisco River Transposition Project
- Urban water use (for human and industrial consumption)
- Rural water use (for human consumption and irrigation)
- Environmental
- Economic and financial
- Drought preparedness measures

The actions were proposed considering that the water resources sector and its institutional framework are primarily responsible for their implementation and/or articulation with other impacted sectors. Table 5.2 presents an example of the character of proposed actions with respect to water allocation and use permits in response to different degrees of drought severity.

TABLE 5.2

Drought Management Thematic Axis Example for the Piranhas-Açu Basin Plan: Allocation of Water and Water Use Permits

Thematic Axis: Water Allocation and Water Use Permits				
Category	**Action**	**Objectives**	**Character of the Actions**	**Responsible Institutions**
D0 Normal	*Traditional* permits + negotiated allocation	Guarantee water quantity and quality for various uses	Voluntary	ANA; SEMARH-RN; AESA-PB
D1 Moderate	Revision of *traditional* permits for large users with small reduction in demand + negotiated allocation	Maximize meeting of demand (considering minimum losses)	Voluntary	ANA; SEMARH-RN; AESA-PB
D2 Severe	Negotiated allocation + climate forecast	Maximize meeting of demand	Obligatory measures	ANA; SEMARH-RN; AESA-PB; CBH-PPA; Management Commissions; Water Users
D3 Extreme	*Hydro-economic* negotiated allocation + climate forecast	Reduce the economic impacts of droughts for all users	Obligatory measures	ANA; SEMARH-RN; AESA-PB; CBH-PPA; Management Commissions; Water Users
D4 Exceptional	*Hydroeconomic* negotiated allocation + climate forecast	Reduce the economic impacts of droughts for all users	Obligatory measures and zero tolerance	ANA; SEMARH-RN; AESA-PB; CBH-PPA; Management Commissions; Water Users

Drought preparedness planning is a process and not a discrete event. After elaborating the protocol for drought preparedness for the Piranhas-Açu River Basin, its implementation will be carefully monitored and the proposed actions revised as appropriate to permit its evolution in response to natural, technological, legal, institutional, political, and management changes, as well as to changes in the basin's needs.

Finally, it is essential to establish strong governance of the protocol through institutional articulation in the basin to ensure *ownership* by stakeholders and create a favorable and collaborative scenario for its implementation.

5.4 Drought Preparedness Plan for a Hydrosystem

Two drought preparedness plans were elaborated for hydrosystems under the technical collaboration program: one for the Jucazinho system in the state of Pernambuco and the other for the Fortaleza metropolitan system in the state of Ceará. Hydrosystems are composed of surface water supply sources (i.e., rivers, lakes, and reservoirs) and/or subterranean ones (i.e., underground and artesian aquifers), for multiple purposes (i.e., human consumption, irrigation, flood control, etc.). A specific hydrosystem can be located in one or more river basins (in the case of river basin transfers). Thus, the boundaries of the hydrosystem are defined in a way that is appropriate for the specific analysis to be undertaken so as to include all water supply and demand sources of relevance for the timescale considered, since the pertinent elements for a long-term analysis can be different from those for a short-term one. Normally, a greater number of elements need to be considered in a longer-term analysis. This system is subject to the legal and institutional framework for water resource management in Brazil, Law 9433 (1997). This law establishes the National Water Resources Policy and the National Water Resource Management System, introducing a new approach to integrated water resource management through the application of planning and economic instruments.

The drought preparedness plan for a hydrosystem aims to define its operating rules in drought conditions of different severities and the water allocation measures associated with each one. The long-term allocation consists of the establishment of water use permits, and the short-term one in use restrictions during a drought event. The operational rules of the system should also include safeguards to avoid extreme drought conditions. These safeguards are associated with average flows in the system or their increase during less severe drought periods.

TRLs in the system should be informed by decision makers. In the Jucazinho case, these levels were defined by the Pernambuco Water and Climate Agency (APAC) and the State Water Supply and Sewerage Company (COMPESA). This structuring of decisions was interactive, in which there was an initial *ad hoc* set of decisions (i.e., based on past experience and not on the use of models) whose possibilities and effects were analyzed through simulations and optimization. This was an iterative and interactive process that evolved during elaboration of different versions of the plan.

The basic structure of operational planning for this hydrosystem included six clusters of activities:

- *Analysis of the legal and institutional context*: (1) Identification of the interested stakeholders and institutions, evaluating their responsibilities and interests; and (2) analysis of current management tools—water rights system, water allocation process, system operation, and so on.

- *Determination of risk tolerance level and the number of drought phases:* This should initially obtained in an *ad hoc* fashion and later revised after an evaluation of the impacts of these decisions.
- *Definition of a drought monitoring system*: Definition of the drought indices for broad research involving meteorological and hydrological indicators; keeping in mind the quantity of false alarms and failures to detect droughts that are actually occurring, the indicator used was reservoir level.
- *Development of simulation and optimization models to analyze drought preparedness and/or response actions*: (1) Supply and demand estimates; and (2) water supply guarantees and risks.
- *Definition of the various alternatives for each stage of the drought or drought category*: (1) Water allocation scenarios; and (2) complementary water sources (including wastewater reuse and desalination, as well as rainfall).
- *Decision making*: (1) Definition of the triggers (target levels of the reservoir or reservoirs); and (2) definition of use restrictions (water supply) for each stage of the drought.

Some implementation measures in the current form of the plan still need to be refined in future iterations, such as inspection mechanisms and compensation for affected stakeholders. In general, however, the actions to be included in the drought preparedness, mitigation, and response strategies should yield four types of results:

- Increased quantity and quality of water supply through the use of existing systems (e.g., water transfers across basins), new systems (desalination, use of subterranean water, etc.), expansion of the conjunctive use of surface and groundwater, or changes in water treatment techniques
- Demand reduction through proactive measures (e.g., legal and economic actions, education, demand prioritization, loss reduction), reactive ones (e.g., wastewater reuse), and adjustments (e.g., urban demand)
- Impact minimization through anticipatory strategies (e.g., regulation, conflict management), loss absorption (e.g., insurance, compensation, reserve funds); and loss reduction (e.g., changes in water use)
- Conflict management through conflict mediation and compensation measures for sectors experiencing greater losses

5.5 Drought Preparedness Plans for Urban Water Utilities

As noted earlier, under the technical collaboration program, drought preparedness planning was carried out for the Jucazinho hydrosystem and for metropolitan Fortaleza. In the case of Jucazinho, the reservoir has two main purposes: urban water supply and flood control. The urban drought preparedness planning exercise for the Jucazinho system will be used to illustrate the methodology applied through the technical collaboration program in Northeast Brazil.

The drought preparedness plan for water utilities is designed purely for the water supply sector, and includes as key actors the water supply utility and the entity responsible for policy and regulation. The main legal framework is the federal water and sanitation law, Law 11,445 (2007). Thus, the system to be analyzed is the water supply system for urban consumption, including drinking water treatment and distribution. The plan focuses mostly on water conservation actions and operational processes of drinking water treatment and water distribution in the city.

The Jucazinho hydrosystem is located in the Northeast portion of the Capibaribe River basin in the *agreste** region of Pernambuco. This system is responsible for supplying water to 15 urban centers of varying population sizes. Its main reservoir, which has the same name, has a maximum capacity of 327,035,818 m^3 and a river basin with a drainage area of 4171 km^2. Thus, it was considered as the principal water source to evaluate water supply in the system.

The methodology applied was that of the aforementioned adaptive and evolutionary planning (see Figure 5.5) in which different methodological configurations for plan elaboration were analyzed. This occurred through a participatory process that included the principal state-level institutional stakeholders of the respective water supply systems (i.e., the state water supply and sanitation company, water agency, and water resources secretariat). The plan was initially constructed through a series of brainstorming meetings with the operators of the water supply system and the agency responsible for water resource management. The first version of the plan reflected a joint vision of the planning process and its challenges. Subsequent versions of the plan made more intensive use of quantitative modeling, which permitted simulation and optimization of the system, together with utilization of robust identification and analysis methodologies for the assessment

* As observed in Chapter 1, the *agreste* is the area located between the humid coastal zone along the eastern coast of the Northeast known as the *zona da mata* and the vast semiarid area further inland, known as the *sertão*. Compared with these two adjacent areas, it is characterized by intermediate levels of precipitation in years of normal rainfall, but can also be adversely affected by drought conditions that primarily affect the *sertão*.

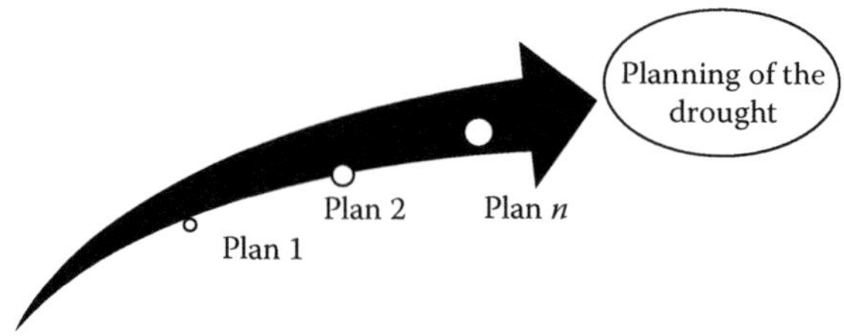

FIGURE 5.5
Drought preparedness planning as an evolutionary process.

of vulnerabilities and the operational definition of monitoring, mitigation, preparedness, and response strategies and actions for droughts. All versions of the plan were elaborated according to the methodological steps indicated in Figure 5.6.

The diagnostic stage of the process sought to identify the legal, institutional, technical, economic, and social conditioning factors, as well as

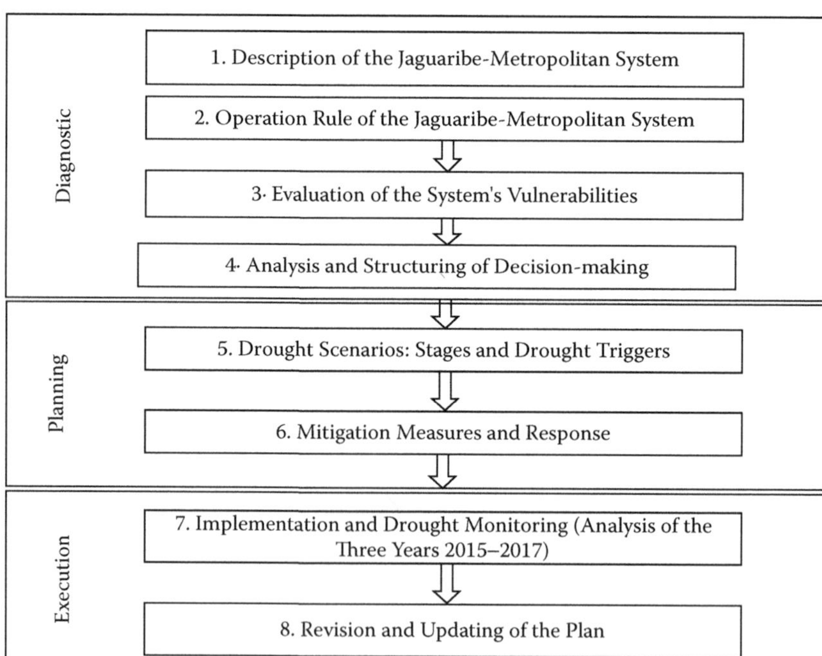

FIGURE 5.6
Methodological steps for the elaboration of an Urban Drought Preparedness Plan. (Adapted from Colorado Water Conservation Board, *Municipal Drought Management Plan Guidance Document*, Colorado Water Conservation Board, Denver, CO, 2010.)

the historical evolution and current status of water supply, with a view toward assessing the vulnerabilities and potential impacts associated with the operational drought. The planning phase, in turn, involved decisions regarding the strategies and actions to be operationalized in order to avoid or reduce the negative impacts of the drought by means of measures designed to reduce the system vulnerabilities identified during the diagnostic stage. Plan execution consisted of monitoring the drought stage and identifying opportunities to implement the strategies and actions proposed during the planning phase. Following each drought event, finally, the experience obtained and lessons learned are expected to be reflected in updating and revision of the plan.

The plan elaboration process took place in a participatory manner over the course of nine months in 2014. During this period, two versions of drought preparedness plans were developed for the Jucazinho system. The strategies and actions that resulted from the planning process defined measures to be carried out prior to the onset of a drought as well as during its incidence. These measures need to be commensurate with the severity of the drought, whose elements can be summarized as follows:

- *Normal (i.e., nondrought period)*: The demands for water are met without any restrictions.
- *Alert*: Administrative and operational preparation for the effective onset of the operational drought.
- *Conservation*: Reducing consumption by means of economic incentives and water conservation campaigns, as well as by increasing financial resources for the expansion of water supply and reducing losses.
- *Restrictions*: Physical reduction of consumption.
- *Emergency*: Severe impacts, measures of high social and economic-financial cost to avoid total collapse of the system.

These measures were classified and an initial identification of actions was made based on the "Guide for Elaboration of Emergency Drought Plans in Urban Water Supply Systems" from Spain (Gonzáles and Morcillo 2007) and the Drought Response Plan for Denver, Colorado (Denver Water 2014). The measures initially identified were widely discussed with the Pernambuco and Ceará state urban water supply system operators and water resource management agencies in order to tailor them to the specificities of each system.

An analysis was next undertaken to provide greater consistency and completeness in terms of the actions proposed. In a first iteration, these actions were organized into the following categories (González and Morcillo 2007):

- *Mitigation measures*: These are intended to help avoid the occurrence of drought impacts (e.g., inventory of alternative sources of supply, monitoring of hydrometeorological parameters).
- *System operation and management measures*: These include the internal actions within the institutions responsible for the operation and management of the public water supply system or systems (e.g., control of losses, rationing).
- *Institutional measures*: This set of actions entails integration of the institutions and agencies with drought management responsibilities (e.g., by disseminating information and establishing dialogue with government authorities, institutions, and users in each area).
- *Legal and normative measures*: These measures should be taken in a way compatible with the legal and normative framework for water resource management (e.g., inspection).
- *Social impact measures*: These refer to actions to reduce water demand and ensure that adequate and useful information is provided to consumers (e.g., provision of information to the media, educational campaigns with the aim of reducing consumption).
- *Environmental impact measures*: Actions intended to reduce the significant environmental impacts from the use of water resources (e.g., control of residual discharges to rivers).
- *Measures to comply with objectives*: These are intended to ensure compliance with the objectives established for each phase of the drought (e.g., increasing financial and human resources to address the drought).
- *Measures to expand and improve water availability infrastructure*: Actions to increase water availability should be included in the drought preparedness plan (e.g., construction of new aqueducts and availability of water tank trucks).
- *Risk monitoring measures*: Actions that seek to develop indicators to assess evolution of the drought situation (e.g., periodic assessment of the situation and development of drought response actions).
- *Preparedness measures*: These are useful for effective implementation of the various actions and have as their objective to prepare plans, projects, or campaigns to manage more severe droughts.

5.6 Final Observations

As noted throughout this book, drought management in Brazil needs to advance from a crisis management approach to one based on risk management. This will result in a reduction of both the costs and the social and economic impacts associated with the drought. Several pioneering drought preparedness plans for water supply and management incorporating a risk management approach have now been elaborated for parts of Northeast Brazil. The next challenge is to refine the methodologies employed and to define public policies that help to better disseminate these innovative drought planning and management practices. Chapter 12 presents in more detail the different steps involved in the participatory elaboration of drought preparedness plans at the level of a river basin, a hydrosystem, and an urban water utility.

6

Municipal Agricultural Drought Preparedness and Response Plan

Barbara Farinelli, Pablo Valdivia, and Diego Arias

CONTENTS

6.1 Background 67
6.2 Characterization of the Municipality 69
6.3 Agricultural Drought Preparedness and Response Plan for Piquet Carneiro 70
6.4 Lessons Learned 76
6.4.1 Identification of Key Stakeholders 76
6.4.2 Effectiveness and Sustainability of the Proposed Plan 76
6.4.3 Strong Leadership Support 77
6.4.4 Managing Expectations and Plan Funding 77
6.4.5 Technical Guidance and Capacity Building 78
6.4.6 Model for Replication of New Plans 78
6.4.7 Information Systems and Continuous Monitoring 78
6.4.8 Inter-Institutional Arrangements and Involvement of the Private Sector and Civil Society 79

6.1 Background

Rain-fed agriculture plays a dominant role in providing food and livelihoods for small-scale farmers in the semiarid region of Northeast Brazil (Rockstrom et al. 2010). However, rain-fed production systems are vulnerable to arid conditions and drought. When extreme drought events occur, vegetative crop cycles are interrupted by severe soil water losses, causing crop failures and negative social impacts on rural populations. During the 2012–2013 drought period, nearly 38% of the Northeast population was affected, corresponding to nine million people (Marengo and Bernasconi 2015).

The government of Brazil has developed several initiatives over the decades to reduce vulnerability of rural populations facing droughts, including water supply by tank trucks, food and seed distribution, and harvest guarantee (*Garantia Safra*) programs (which are targeted to poor farmers and cyclical with the crop season) (Larreguy and Monteiro 2013). As discussed in previous chapters, however, this reactive crisis management strategy needs to evolve as governments recognize the importance of developing public policies oriented to strengthening resilience to climate change, boosting natural resource conservation, adopting appropriate economic development strategies, improving access to social services, supporting the poor in less developed areas, spreading knowledge, and helping people to take ownership of these strategies and policies.

As part of the technical collaboration program to support the Brazilian federal government to help develop an integrated dialogue and strategy for national and regional drought management, a drought preparedness and response plan for rain-fed agriculture was developed to demonstrate tangible strategies and tools for proactive drought management at the municipal level. The municipality of Piquet Carneiro, located in the state of Ceará, was selected as a pilot for the new approach based on the following criteria: (1) a low 2010 Municipal Development Index* (Índice de Desenvolvimento Municipal—IDM) value; (2) medium-high to high vulnerability to droughts according to the 2013 Municipal Alert Index (Índice Municipal de Alerta—IMA)†; and (3) strong institutional coordination and active social groups, including producers' associations.

The plan was developed through a participatory process with key stakeholders, including groups of agricultural producers, their associations and unions, representatives of Piquet Carneiro's municipal administration, the municipal association, the Ceará Drought Committee, the Secretariat of Agrarian Development, and members of the rural technical assistance and extension services. The methodology used for its elaboration included a review of legal, institutional, and operational aspects related to drought risk management at the federal, state, and municipal levels. Four strategic lines were recommended in the plan for institutional strengthening, adoption of management tools, training and capacity building, and infrastructure investments to provide effective drought risk management for the municipality.

---

* The Municipal Development Index (IDM) is composed of indicators in the following areas: (1) physiographic, land, and agriculture; (2) demographic and economic; (3) supporting infrastructure; and (4) social. It provides relative values for 184 municipalities in Ceará. The IDM values considered for purposes of the pilot were those for municipalities with the lowest levels of development. This category groups the 105 least developed municipalities in Ceará, which together contain 25.5% of the state's population. For more on the IDM, see http://www.ipece.ce.gov.br/categoria4/idm/IDM_2010_valores.pdf. Accessed on May 3, 2014.

† The Municipal Alert Index (IMA) is based on the analysis of 12 indicators that together categorize the degree of vulnerability of municipalities on four different dimensions. The purposes of this indicator are to (1) identify and rank the municipalities with greater vulnerability; (2) select areas with the greatest likelihood of social tensions; and (3) define criteria for selection of the municipalities that will benefit from emergency, mitigation, or structural actions.

Due to the success of plan preparation, the state government is considering the institutionalization of the Piquet Carneiro approach in order to replicate it in other municipalities where agriculture production is also predominantly rain-fed. The major challenges now are to ensure that (1) plan implementation and sustainability involve the appropriate alignment of federal, state, and municipal policies; (2) institutional capacities and arrangements are strengthened; (3) the necessary financial resources are allocated; and (4) local communities are engaged.

Including this brief introduction, this chapter is organized in three sections with the objective of presenting the overall development of the Piquet Carneiro Plan. The next section summarizes the methodology for plan development and implementation. The subsequent section identifies lessons learned and provides recommendations on some practical and operational activities for the plan.

6.2 Characterization of the Municipality

Piquet Carneiro presents similar characteristics as—and challenges faced by—numerous other municipalities in the Northeast semiarid region. The municipality is located in the center of the state, and has an area of 587.8 km^2. It has a hot tropical semiarid climate, with average temperatures ranging from 26°C to 28°C. Average rainfall in nondrought years reaches 897.6 mm, and is concentrated in the period from February to April. In 2014, the accumulated rainfall from January to October reached 671.3 mm, while, in 2013, the annual total was only 477 mm. Piquet Carneiro is part of the Banabuiú River Basin and its main reservoirs are Ema dos Marinheiros, Timbaúba, Açude dos Macacos, Açude Velho, and São José II. The latter, with a storage capacity of 29.14 million cubic meters, is the main source of urban water supply for the municipality.

The estimated population of Piquet Carneiro in 2015 was 16,461 inhabitants, with nearly 52% living in rural areas. With a GDP of US$17.65 million, the economy of Piquet Carneiro relies predominantly on smallholder agricultural production. In 2013, the main temporary crops in terms of harvested area were corn (2444 ha) and beans (2289 ha), followed by castor beans (55 ha), rice (50 ha), herbaceous cotton (40 ha), and sugarcane (10 ha).

As with many other municipalities in the Northeast semiarid region, Piquet Carneiro is regularly affected by drought events. Analysis of IMA's records for 2004–2013 revealed that the municipality reached medium to high levels of vulnerability in 8 of these 10 years. The highest vulnerability levels were registered for 2012–2013, which experienced one of the worst years on record. Crop yields fluctuated considerably due to erratic rainfall pattern. In Piquet Carneiro, extreme negative deviations in normal crop yields are recorded twice every 10 years. It is estimated that the average value of annual production losses in the main food and cash crops is US$124,820 and the maximum loss is close to US$485,670. On average, beans account for the highest production losses (US$56,420), followed by corn (US$49,185) and cotton (US$19,242). These values are commensurate with these crops' levels of resistance to stress caused by water deficit.

6.3 Agricultural Drought Preparedness and Response Plan for Piquet Carneiro

The extreme drought conditions recorded in Ceará from 2012 to the present, together with the vulnerability of Piquet Carneiro's agricultural sector to this type of event, generated an appropriate context for the implementation of an agricultural preparedness plan for drought management.

The plan was developed through a participatory process that included consultations, workshops, interviews, and meetings with key partners. Plan preparation involved several phases, including the following (see Figure 6.1):

- *Diagnosis*: Definition of a methodological process, data collection, and analysis of policies, the legal framework, and agro-meteorological information.
- *Planning*: Identification of strategic priorities, programs, and projects, including estimated budgets, and definition of dissemination steps.
- *Development*: Alignment of existing and definition of new programs and policies, including public–private partnerships to implement the plan, discussion of the plan's institutional framework with the main stakeholders, and definition of a logical framework (or logframe) of plan activities.
- *Dissemination*: Preparation of a dissemination plan and of awareness measures for the plan, development of practical guidelines, identification of financing sources and estimated costs for effective plan implementation, and provision of advice to the Drought Committee of Ceará for effective plan implementation, with particular emphasis on mitigation of drought risk and emergency response.

The plan's general objectives were to (1) reduce the vulnerability of farmers to extreme drought events and (2) improve existing coordination mechanisms among municipal and state institutions. Its strategic priority was to improve current drought management activities in the municipality. (Details of the plan are shown in Box 6.1.) Consequently, it sought to promote institutional

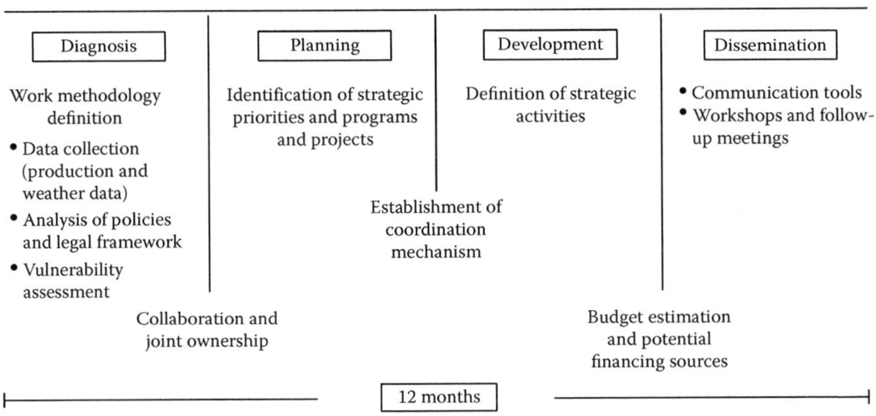

FIGURE 6.1
Timeline for preparation of the municipal agricultural drought preparedness and response plan for Piquet Carneiro.

BOX. 6.1 MUNICIPAL AGRICULTURAL DROUGHT PREPAREDNESS AND RESPONSE PLAN IN PIQUET CARNEIRO

The Municipal Agricultural Drought Preparedness and Response Plan aims to reduce the vulnerability of farmers to extreme drought events. This plan targets around 2800 family farmers who have an annual gross income between US$2,600 and US$93,000.

The plan recommends specific activities for the following strategic intervention areas:

Strategic Line of Intervention No. 1 (LE1) aims to strengthen local institutions, particularly Municipal Civil Defense Coordination Unit (COMDEC). In addition, the activities under LE1 seek to create proper conditions for COMDEC to play a leading role in plan implementation. This line of intervention was recommended based on the assumption that strengthened institutions and definition of clear roles and responsibilities among different stakeholders are needed for effective coordination and implementation of the plan.

The activities under Strategic Line of Intervention No. 2 (LE2) were recommended with the objective of supporting plan implementation through the development of (1) information tools (i.e., farmers' database and weather information), (2) coordination mechanisms and cooperation agreements, and finally, (3) a

(Continued)

communications strategy. Implementation of all LE2 activities will improve the decision-making process for local authorities and farmers, thereby reducing the likelihood of adverse effects on crop and livestock production systems.

Similarly, the objective of Strategic Line of Intervention No. 3 (LE3) is to build technical capacity among different stakeholders, including both producers and rural extension officers, in Piquet Carneiro. Recommended activities include increasing the volume and availability of information on drought risk management for extension officers and producers and enhancing the understanding of drought conditions to help induce adoption of better production technologies.

Finally, Strategic Line of Intervention No. 4 (LE4) recommends investing in infrastructure (i.e., building water storage facilities and more efficient irrigation systems, among others).

The estimated cost for initially implementing the action plan is US$3.75 million, with US$3.37 million allocated to LE4, followed by LE3 with US$0.35 million, and LE2 and LE1 with US$0.02 and US$0.01 million, respectively.

strengthening, adoption of management tools, training and capacity building, and infrastructure investments to support adequate drought risk management. Specifically, its desired outcomes were as follows:

- The COMDEC has a clear view of the coordination of drought preparedness and response efforts.
- Instruments and processes for drought preparedness and response are reinforced.
- Drought coping capabilities of producers are improved.
- Infrastructure for dealing with drought conditions is strengthened.

Given the socioeconomic importance of the agricultural sector in Piquet Carneiro, key stakeholders, including from both state and municipal organizations, worked in close collaboration to develop the plan. During its preparation over a 12-month period, a set of key tools was designed or adapted to accomplish the following tasks.

Assess existing laws and coordination mechanisms for disaster risk management activities at the municipal, state, and national levels. Varvasovszky and Brugha (2000) stress that a stakeholder assessment can help understand not only the evolution of policies, but also whether it is feasible to implement new policy directions. In addition, valuable information is obtained to evaluate the challenges inherent in plan implementation and develop management strategies for key stakeholders.

Analysis of the legal framework and the stakeholder assessment should include, *inter alia,* a review of existing legislation relating to disaster risk management, political priorities and sectoral targets, institutions (national, state, municipal, etc.) that provide support to beneficiaries, description of the decision-making process for the allocation of resources for drought preparedness and response activities, as well as provision of information for and design of a monitoring and evaluation (M&E) system.

In the case of Piquet Carneiro, the analysis found that the state constitution encouraged cooperation among federal, state, and municipal institutions with the aim of promoting socioeconomic development. To achieve this goal and avoid duplication of efforts, the state government created several institutions and mechanisms, including the State Council for Permanent Actions against the Drought, the Integrated Committee to Combat Drought (Comitê Integrado de Combate à Seca), and the State Civil Defense System, as well as local structures such as COMDEC.

Even though the state constitution mandates all public organizations in Ceará to carry out drought preparedness efforts, they normally invest greater resources for emergency response activities than for the adoption and implementation of a comprehensive *ex ante* risk management strategy. In Piquet Carneiro, for instance, COMDEC is responsible for the development of damage and loss assessments at the district level and then compiling this information for the municipality as a whole. This information is then shared with the Regional Civil Defense Coordination Units (COREDECs) and State Civil Defense Coordination Unit (CEDEC) for decision and response support.

Reducing the vulnerability of Piquet Carneiro, or that of any other municipality in Brazil's semiarid region, to drought events is subject to the existing risk management frameworks at both state and federal levels. The legal framework for risk management at these levels must support concrete actions concerning mitigation (e.g., adoption of sustainable agricultural practices, water management infrastructure), transfer (e.g., purchasing insurance, financial hedging tools), and coping/social protection efforts (e.g., government conditional or unconditional cash transfer programs, buffer funds).

Illustrate linkages between issues that prevent key institutions from conducting drought response and mitigation activities (i.e., problem tree analysis). This process leads to a better understanding about "the interrelationship of problems and opportunities, strengths, weaknesses, and threats" (Wisner 2006, p.323), and potential impacts when conducting projects or programs with specific objectives. See Figure 6.2 for the problem tree analysis for Piquet Carneiro.

The self-assessment carried out by relevant stakeholders, including representatives of the municipal council of Piquet Carneiro, the Agrarian Development Secretariat of Ceará (SDA), the Association of Municipalities of Ceará (APRECE), the Technical Assistance and Rural Extension Enterprise of Ceará (EMATERCE), and the COMDEC of Piquet Carneiro, revealed a

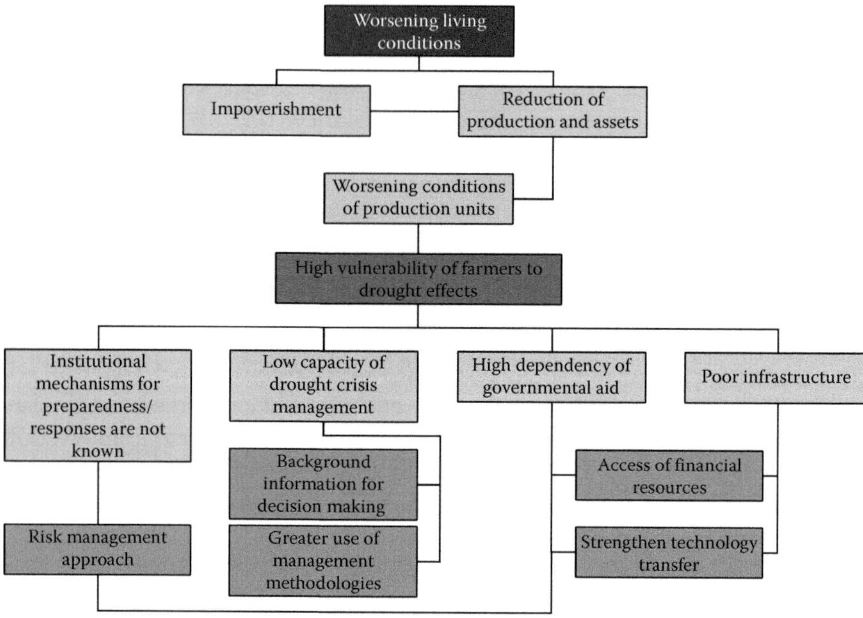

FIGURE 6.2
Problem tree analysis, defined by the stakeholders in Piquet Carneiro during the drought planning process for the reduction of farmer vulnerability to the effects of extreme droughts.

number of challenges in terms of reducing farmer vulnerability to drought events. For instance, there is a lack of coordination among government institutions and civil society due to the absence of a specific legal framework for drought risk management in agriculture. As mentioned above, the available financial resources are mainly allocated for emergency relief activities, most of which are conducted without the guidance of a plan or strategy. As a result, the long-term positive impact of these actions on vulnerable groups is generally limited, thus increasing the likelihood that emergency situations will be repeated in the future.

Similarly, the provision of technical assistance and rural extension support to farmers has been considered by local authorities in Piquet Carneiro to be a core factor to ensure that *ex ante* risk management activities are implemented. To this end, the annual budget of the Municipal Secretariat of Family Farming allocates resources to increase the number of small producers that receive extension services. Despite the significant efforts of local authorities to decrease farmer vulnerability, financial resources at the municipal level are insufficient to adequately fund a comprehensive drought risk management plan. Consequently, implementation of the drought management plan needs to rely almost entirely on federal or state funds.

Assess the capability of coping with drought events (i.e., assessment of institutional capacities). The objective of this analysis is to help local institutions better understand their capacities and weaknesses. Through a participatory process, stakeholders can identify the resources available to cope with risks. Lebel et al. (2006) have proposed a comprehensive framework that can be used to assess institutional capacities and practices before, during, and after a disaster. Their approach focuses on four dimensions: (1) the capacity to deliberate, which reflects whether the interests of all stakeholders are represented; (2) the ability to access and coordinate the disbursement of resources during the various phases of a disaster (i.e., preparedness/mitigation, emergency, and recovery); (3) the capacity to implement risk management activities*; and (4) the capacity to evaluate risk management activities (which is important because the information derived from this analysis may identify opportunities for improvement).

Assess direct drought impacts on the agricultural sector.† Impact data is critical for understanding drought vulnerabilities,‡ including identifying the worst event ever recorded. Key parts of this analysis are the frequency of the occurrence of drought events and the estimation of fiscal expenditures allocated for recovery activities.

Given the direct relationship between soil water availability and crop yields, drought impacts are relatively easy to observe in the agricultural sector. The occurrence of this type of weather event causes negative supply shocks.§ To simplify calculation of the effects of droughts, the process was limited to direct crop losses for the Piquet Carneiro Plan.¶ A loss is considered to have occurred when a current yield is less than 70% of the normal one. This threshold helps to distinguish between losses that are triggered by significant shocks and those that reflect absorbable downturns in a municipality like Piquet Carneiro. Results obtained during this process were then correlated with weather indicators (i.e., Standardized Precipitation Index—SPI).**

* As part of this analysis, it should be determined whether both public and private institutions are well prepared and informed. In addition, it should be determined whether the executed activities help reduce the likelihood of a severe impacts from a future drought event.

† The National Drought Mitigation Center (NDMC, n.d.) has developed a comprehensive list of drought impacts that can be used by planners as a reference for undertaking impact studies. The drought impact analysis should be conditioned upon the time and resources available.

‡ Ding et al. 2011; Wilhite, Svoboda, and Hayes 2007 cited in Lackstrom et al. (2013).

§ The economic impacts of droughts are not always negative. For instance, producers from a region that have a surplus of certain goods will be motivated to sell their products into the drought-affected market because of favorable prices (Ding et al. 2011).

¶ Drought can cause long-term impacts on livestock production systems. To estimate negative impacts in this sector, it is advisable to set a timeline for the analysis.

** In Brazil, as mentioned in Chapter 4, national and regional meteorological and hydrological services generate a variety of meteorological and agricultural drought indices that can be used to characterize drought events. A survey conducted by Sentelhas (2010) identified the most common indexes generated in Brazil as Rainfall Anomaly, Standardized Precipitation Index (SPI), Palmer Drought Severity Index (PDSI), Number of Days Without Rain (NDWR), Accumulated Drought Index (ADI), Accumulated Water Deficit Index (AWD), Relative Water Deficit Index (RWD), Crop Moisture Index (CMI), Water Requirement Satisfaction Index (WRSI), Soil Water Storage (SWS), and Normalized Difference Vegetation Index (NDVI).

Similarly, historical datasets were reconstructed for direct losses or damages in the agriculture and livestock sector based on statewide damage and loss evaluation reports and government emergency decrees.

6.4 Lessons Learned

The team responsible for carrying out the plan preparation for Piquet Carneiro identified several key lessons regarding the factors required to improve its governance, implementation, and sustainability. These are briefly described in Sections 6.4.1 through 6.4.8.

6.4.1 Identification of Key Stakeholders

The identification and engagement of stakeholders are important steps to (1) build awareness of the approach, its objectives, and opportunities; (2) minimize the likelihood of plan failure; and (3) generate political ownership. This process helped the plan preparation team, which comprised consultants, representatives from the municipal council, SDA, and EMATERCE, to identify the main problems that hinder efficient and effective drought risk management in the municipal agricultural sector.

The preparation team conducted a broad consultation process, including discussions for the development of the plan proposal and intermediate validations with different stakeholders in the agricultural sector, including groups of agricultural producers and state institutions (i.e., SDA, EMATERCE, APRECE, the Drought Committee, and the World Bank-financed Ceará Rural Sustainable Development and Competitiveness Project—PSJIII*).

6.4.2 Effectiveness and Sustainability of the Proposed Plan

To increase the effectiveness and sustainability of plan execution, it is crucial to coordinate the key government institutions and to create linkages to existing policies and programs. The involvement of different stakeholders will help to promote harmonization of government actions and policies for drought management and avoid duplication with related programs. To this end, the operation of 12 programs (e.g., *Garantia Safra*) funded by

* The São Jose III Project is the third generation of a community-driven rural poverty reduction initiative in the state that finances investment operations. It provides funds to the family farming sector for the implementation of both agricultural and agroindustrial projects that are linked to beekeeping, aquaculture, staple, sheep, and goat production systems. The project contributes to the provision of resources in the areas of technology, irrigation, and support to small producers to enable them to reach new technological levels and better product quality.

the federal and state governments are implemented by extension officers from EMATERCE. The monitoring process of these programs is carried out through a centralized database system, which uses the required Cadaster of Individuals (CPF) as the basis for registering program activities and linking them with beneficiaries' basic information data (i.e., name, address, and others).

EMATERCE's monitoring system enables government authorities to assess whether national or state programs are adequately targeting beneficiaries, or if beneficiaries are receiving a comprehensive type of assistance (i.e., access to credit and technical assistance, risk transfer mechanisms, and market information). During plan preparation, however, a review of all programs implemented in Piquet Carneiro showed the need to improve some of the operational processes of the system. For instance, it was found that only around 40% of all CPFs (or 3405 out of 8566 such documents) were properly recorded. As a result, it was difficult to determine the number of farmers being assisted and whether a concentration of such programs existed in specific geographic areas.

6.4.3 Strong Leadership Support

Effective implementation of drought risk management relies on institutions with clearly defined roles for plan coordination and implementation. During plan preparation, COMDEC was identified as the responsible institution for conducting these activities. Thus, it is necessary for COMDEC to be recognized as plan coordinator, understand its role, and be able to perform this function, thereby creating the basis for successful plan implementation. Regulating the law is the most straightforward way to authorize COMDEC to lead this work because the regulation provides details to individuals, institutions, nongovernmental organizations, and others about their roles, responsibilities, and what penalties will apply if there is failure to comply with the law. However, despite the present absence of these regulations in Piquet Carneiro, members of COMDEC's operational structure (i.e., the Municipal Secretariat of Family Farming, Municipal Secretariat of Environment, etc.) successfully supported and followed the guidelines proposed by its president.

6.4.4 Managing Expectations and Plan Funding

The application of integrated management approaches requires behavioral and technical changes in the traditional culture at all levels of government, as well as among beneficiaries. On the technical side, for example, it requires mapping all activities necessary for plan implementation, estimating costs, identifying different funding sources, and determining and acting upon potential "quick wins."

6.4.5 Technical Guidance and Capacity Building

Effective implementation of drought preparedness and response activities depends upon an active learning process to strengthen the capacity of the stakeholders involved in order to reduce their vulnerability. One of the plan's proposed strategies aims to strengthen the skills of producers, rural technicians, and other relevant groups in order to expand their knowledge, enable access to information, boost understanding, provide appropriate new experiences for coping with the drought, increase their ability to innovate, and better adapt to changing production conditions in the semiarid region. During plan preparation, the technical knowledge gap on the part of producers and agricultural technicians (i.e., extension workers) that directly contributes to low levels of appropriate agricultural production technology use was identified.

6.4.6 Model for Replication of New Plans

The preparation team in Piquet Carneiro estimated that the cost and timeline for the elaboration of new plans can be significantly reduced if a slightly different approach is implemented. Future plans should be prepared by local consultants that lead the technical work in neighboring municipalities. For instance, personnel from EMATERCE, SDA, and representatives of COMDEC in Piquet Carneiro could help the nearby municipalities of Senador Pompeu, Irapuan Pinheiro, Acopiara, and Mombaça draft their plans. Similarly to the case of Piquet Carneiro, state authorities should identify one or two leading municipalities per region to perform the work described above.

Finally, the calculation and interpretation of agrometeorological indices can be outsourced to local (i.e., FUNCEME) or regional meteorological institutions (i.e., the Northeast Drought Monitor). Should the recommendations outlined above be followed, it is estimated that the timeline for preparation of new plans would be reduced from 12 to 6 months and the total preparation cost would decrease from US$100,000 to US$20,000 per municipality.

6.4.7 Information Systems and Continuous Monitoring

The decision-making process for agricultural drought preparedness and response activities must be supported by an objective, transparent, and reliable drought monitoring system. Specialized meteorological agencies in Northeast Brazil, in collaboration with the National Water Agency (ANA) and the World Bank, are carrying out significant efforts in the design of the Northeast Drought Monitor as summarized in Chapter 4. The information provided by this tool can help decision makers make more efficient and effective use of limited financial resources, as they can be better directed to the localities and the people most in need.

Currently, the Monitor computes indices to characterize the severity of drought conditions throughout the region. However, it is soon expected also to address challenges linked to the agricultural sector, which requires information with a higher degree of spatial resolution and greater temporal frequency. The information derived from the Monitor will provide a valuable input to estimate crop yield responses to water deficits and can potentially be used to support critical short-term operational decisions (e.g., to decide planting dates or extend the grazing period).

6.4.8 Inter-Institutional Arrangements and Involvement of the Private Sector and Civil Society

The absence of a drought policy in a country generally means that national and subnational institutions conduct fragmented activities, mostly with limited or no coordination (Bazza 2001). Therefore, a crucial element in the plan implementation process is that it facilitates harmonized efforts for drought risk management, including in the agriculture sector. Coordination and cooperation mechanisms between public and private institutions are crucial for improving efficiency and effectiveness in the use of resources. The increased incidence of natural events raises the cost of disaster risk management activities for the public sector. However, private organizations can help reduce costs when contributing resources and knowledge for implementation of risk preparedness and response efforts. In the same way, establishment of public–private partnerships for disaster risk management can improve the capacity of communities to be prepared and respond to shocks due to extreme droughts. To this end, the United Nations Office for Disaster Risk Reduction (UNISDR 2004) identifies four areas of opportunity for the private sector to better manage risks, specifically, monitoring hazards and communicating risks, sociophysical strengthening, sharing financial risk, and disaster preparedness.

An organizational structure and social leadership similar to those which currently exist in Piquet Carneiro, helps to strengthen efforts and avoid duplication of work. Similarly, the involvement of civil society is essential to reducing the community's vulnerability to shocks through its assistance in the identification and implementation of risk management strategies.

A critical factor in the success of the implementation process of the Piquet Carneiro plan has been the unconditional support and leadership of the SDA of Ceará and of the municipal government. Local ownership of the process is crucially important to ensure sustainability of the plan (FAO and NDMC 2008). The main representatives of the municipal council of Piquet Carneiro were actively engaged in plan preparation, and are now actively collaborating with SDA personnel to develop similar tools, first in the neighboring municipalities of Deputado Irapuan Pinheiro and Milhã, and then for the rest of Ceará.

For the time being, the government of Ceará is funding the short-term strategic activities of the plan. For instance, the São Jose III Project is tendering a contract to install agrometeorological weather stations to monitor drought events in Piquet Carneiro. Several workshops are also being hosted in the municipality with the aim of disseminating plan activities among representatives of local institutions and small producers. In addition, a new center for the provision of rural extension services has recently been approved for establishment in Piquet Carneiro to deliver technical assistance to producers on sustainable and more drought-resilient agricultural practices.

7

Perspectives from the Outside: Contributions to the Drought Paradigm Shift in Brazil from Spain, Mexico, and the United States

Michael Hayes, Mario López Pérez, Joaquín Andreu, Mark Svoboda, Brian Fuchs, Felipe Arreguín Cortés, and Nathan L. Engle

CONTENTS

7.1 Introduction 81
7.2 Overviews of the Processes in Spain, Mexico, and the United States 82
7.2.1 Spain 82
7.2.2 Mexico 83
7.2.3 The United States 84
7.3 Supporting the Brazilian Paradigm Shift 85
7.4 Lessons and Recommendations from These Experiences 89
7.5 Conclusions 90

7.1 Introduction

The progress exhibited by Brazil in tackling the drought could not have been achieved without a strong international push to reform drought approaches to be more proactive. Nor could it have happened without significant support from other countries. As noted in Chapter 3, the World Bank played a key convening role in identifying and mobilizing exchanges to share experiences and learning between policy officials and technical experts in Brazil with key counterparts in other countries. These were mainly from Spain (numerous local government and academic experts), Mexico (CONAGUA, the national water authority), and the United States (the National Drought Mitigation Center, or NDMC, various government officials including the National Integrated Drought Information System [NIDIS], NGO representatives, and academic experts).

To catalyze these exchanges, the World Bank first investigated drought policy and management across a handful of countries to compare approaches and identify common lessons and recommendations. It subsequently brought representatives from the above-mentioned institutions to Brazil for a series of workshops over a two-year period to share lessons and help teach the Brazilian partners the nuances of their countries' approaches. The idea was that Brazil would adapt and build from these various processes, products, methodologies, and tools to fit its unique national context. In addition, the World Bank designed a learning exchange between Brazil, the United States, and Mexico, through which four Brazilian professionals spent a week each in the two latter countries learning how to develop drought monitoring and early warning systems (DEWS) and drought preparedness plans. The World Bank also sponsored two technical study tours, one to the United States and the other to Spain, which involved over a dozen Brazilian officials responsible for water and drought policy and management and their respective counterparts in those countries.

This chapter provides a more detailed account of these exchanges between Brazil and Spain, Mexico, and the United States. It does so from the perspective of several of the key country partners who have collaborated to produce this chapter. The chapter offers insights that situate the recent experience in Brazil among efforts in these three countries and with respect to the broader drought policy shift occurring around the world. It also provides recommendations, from an external perspective, on the next steps Brazil might take as the paradigm shift continues in the Northeast region and beyond.

7.2 Overviews of the Processes in Spain, Mexico, and the United States

7.2.1 Spain

Many river basins in eastern and southeastern Spain share common characteristics related to aridity, water scarcity, and high hydrological variability. As a result, these basins are prone to frequent and intense droughts, while also exhibiting a long history and tradition of adaptation to them. Examples of these technical and institutional adaptation strategies include irrigation systems, ditches and reservoirs, wells, water transfers, desalination plants, water tribunals, and river basin partnerships.

During the 1980s, efforts to implement long-term planning at the river basin-scale began and, by 2007, all river basins across Spain had drought management plans. These experiences encouraged the engagement of IIAMA (the Institute for Water and Environmental Engineering) of the Universidad Politécnica de Valencia (UPV) to support the Brazilian technical collaboration program. IIAMA performs multidisciplinary and applied research and innovation in many areas related to water resources, from hydrology and hydraulics to water engineering, environmental engineering, water microbiology, and environmental impact assessments. Even though it was officially created only in 2000, many of the research groups involved have been active since the late 1970s and early 1980s. Therefore, there is a long tradition of interaction with water-related institutions and enterprises. The Water Resources Engineering Research Group (WRERG) is one of these, and has produced methodologies and tools that have been very helpful in the analysis of the basins, assisting the development of basin plans, and for the management and operation of water systems.

The Jucar River Basin, which flows to the Mediterranean Sea in eastern Spain, is managed by the Jucar District Partnership (Hydrographic Confederation of the Jucar, CHJ), which oversees an area of 42,989 km^2, including several adjacent basins. The use of models and decision support systems (DSSs) has played an important role in the development of the CHJ Basin Plans for nearly three decades, as well as in the development and implementation of special drought plans (SDPs, i.e., drought preparedness plans). The SDPs have been formulated according to a proactive approach to drought preparedness and mitigation. They include long- (planning), medium- (alert), and short-term (emergency and mitigation) measures that are activated using standardized operative drought monitoring indicators (SODMI) obtained from particular combinations of data on precipitation, reservoir storage, groundwater levels, and river flows collected by an automatic data acquisition system. The SODMIs and threshold curves for assessment of the drought situation have been calibrated by intensive use of DSSs for drought-risk estimation. In addition to their use in the development of the threshold curves used with SODMI, DSSs are also regularly used for real-time management at board meetings to assess drought risks and vulnerability over short- and medium-term time horizons, ranging from a few months to an entire hydrological year (ranging from October to September), or even two hydrological years.

During the hydrological year 2004/2005, a severe meteorological drought led to an intense hydrological drought within the Jucar River Basin. As a result, a Permanent Drought Committee (PDC), with special powers to administer the basins of CHJ under emergency situations was established. The PDC comprised representatives of the Ministry of Agriculture; the CHJ; regional governments; the agricultural, industrial, and urban users; the Spanish Geological Institute; labor unions; and nongovernmental environmental organizations. Its mission included making decisions on water management during the drought, performing continuous monitoring in order to control the efficacy of decisions, following the evolution of drought events and their impacts, and authorizing emergency activities. Implementation of the PDC became a very useful experience for other activities, such as review of the operating rules, design of SDP, and preparation of subsequent versions of the basin plan.

7.2.2 Mexico

Prone to frequent and intense droughts since the pre-colonial times, Mexico's traditionally reactive approach to drought risk management was severely challenged during the 2010–2012 drought that affected most of the country. This experience was the catalyst for the development of a proactive approach to prepare for future drought events.

These recent efforts are being led by the Mexican Water Authority (CONAGUA), created in 1989, and the Mexican Institute of Water Technology (IMTA), established in 1996. In 2013, CONAGUA was appointed by the Mexican President to coordinate and guide the National Program against Drought (PRONACOSE). IMTA was requested to assist CONAGUA to implement PRONACOSE in developing Drought Prevention and Mitigation Programs (PMPMS) with some river basin councils, training other institutions in the process of its elaboration, and implementing other activities, research, and projects on behalf of CONAGUA. CONAGUA is not only the federal authority responsible for water resources management, but also constructs all federal water infrastructure (i.e., for irrigation, water supply, and flood control), enforces the water law and grants water rights, conducts the water rights registry, promotes the provision of water services, develops and coordinates the national water planning process, and collects water fees from users. IMTA carries out multidisciplinary and applied research and innovation in hydrology, hydraulics, irrigation, water engineering, environmental engineering, water quality, and environmental impact assessment.

In addition, Mexico, through CONAGUA, began collaborating with Canada and the United States in 2002 to produce the monthly North American Drought Monitor (NADM) assessment product. Mexico has also studied the experiences of Spain, several states in the United States, Australia, India, and China to develop national drought prevention and mitigation strategies (including the PMPMS). Particular attention was placed on drought planning experiences in California and Colorado, as well as on the existing policies in Texas for some of its public water systems. Mexico has also collaborated on several activities with international organizations, particularly the World Meteorological Organization (WMO), for training and skills development. This has included partnering with the Integrated Drought Management Program (IDMP), which was launched during the High Level Meeting on National Drought Policy (HMNDP) in Geneva, Switzerland, in 2013, and using related HMNDP and IDMP documents as references in the ongoing efforts to define the Mexican National Drought Policy. Recognizing the value of these experiences led the World Bank to approach Mexico for assistance in Brazil.

7.2.3 The United States

In the United States, the NDMC was established in 1995 around the drought management program built by Dr. Donald Wilhite from the early 1980s through the mid-1990s. The NDMC's mission emphasizes a combination of research, outreach, and operational activities that stress a proactive risk management approach to prepare for drought events in contrast to the reactive crisis management approach traditionally followed by officials responding to disasters. NDMC is recognized nationally and internationally for the quality of its drought risk management programs, which are focused on drought monitoring and early warning information systems, planning, and mitigation (i.e., actions taken to reduce drought risks before an event occurs).

As an organization that works at the interface between scientists and policy makers, decision makers, and the public, NDMC now has 18 staff members with very diverse backgrounds in both the physical and social sciences. This mix of expertise has helped NDMC to address many of the complex issues and needs that exist in relation to drought. It also allows NDMC to conduct both dedicated basic and applied drought-related research together with the translation and transferring of that knowledge and processes into operational tools, stakeholder engagement, and educational activities.

Over the years, NDMC staff have participated in multiple US national drought policy institutions including the Western Drought Coordination Council, the National Drought Policy Commission, the National Drought Resilience Partnership, and the NIDIS. NDMC, in fact, was a key participant in the development of NIDIS, a multiagency effort to help coordinate and communicate drought early warning information between federal, state, tribal, and local officials. NDMC also serves on the advisory panel for several international initiatives, including the aforementioned IDMP, which is being jointly led and implemented by the WMO and Global Water Partnership (GWP).

Because of its national and international drought risk management expertise and experience, the World Bank engaged NDMC as a partner in the Brazil project. In particular, the technical collaboration program showed interest in the US Drought Monitor (USDM; a weekly drought monitoring assessment tool for the United States) process. USDM's process involves a partnership between the NDMC, the National Oceanic and Atmospheric Administration, and the US Department of Agriculture, as well as sustained involvement from approximately 370 experts from around the country who provide inputs and validation for the process each week. The particular focus of the NDMC's efforts in the technical collaboration program has involved design of a comprehensive process to monitor droughts in the nine states of Northeast Brazil (i.e., the Northeast Drought Monitor), similar to that implemented via the weekly USDM (http://droughtmonitor.unl.edu).

7.3 Supporting the Brazilian Paradigm Shift

The drought risk management experience within the Jucar River Basin in Spain and its potential applicability to the Brazilian context prompted the World Bank to contact IIAMA-UPV for support. Joaquín Andreu, Abel Solera, and Javier Paredes participated in workshops organized by the World Bank and various Brazilian institutions as part of the technical cooperation project "Water Resources Planning and Adaptation to Climate Variability and Climate Change in Selected River Basins in Northeast Brazil" in 2011, 2012, and 2013. Joaquín Andreu was then invited by the World Bank to participate in the "Adaptation Futures 2014 High Level Workshop on Drought Policies" meeting in Fortaleza, Ceará, in May 2014. At this meeting, Andreu made presentations on "Spanish Drought Policy" and "Drought Management Experiences." IIAMA-UPV also hosted a delegation of Brazilian professionals involved in the World Bank project in September 2014, and presented its experience in several aspects of water research and applied research and innovation. The WRERG at IIAMA-UPV organized the International Conference on Drought Research and Science Policy Interface in Valencia in March 2015, which included a special session on the activities of the World Bank related to drought, with an emphasis on the Brazilian process.

Likewise, since the capabilities and experiences of Mexico (i.e., in terms of the number of professionals, information availability, and institutional arrangements) for the development and implementation of PRONACOSE are similar to those of Brazil, the World Bank was very interested in involving Mexico to assist the Brazilian efforts. As with the World Bank and Brazil, the WMO, Turkey, and other Latin American countries have also been interested in the Mexican experience, and its experts have attended several regional and international workshops to explain the principles and progress of PRONACOSE and offered technical assistance.

As a result, CONAGUA offered training for the Brazilian professionals in March 2014, which focused mainly on Mexico's experiences in drought monitoring and preparedness plan development. Mexico illustrated, in detail, both the process and inputs for the Mexican Drought Monitor's preparation and publication, as well as for the development of the PMPMS (i.e., drought preparedness plans) and the PRONACOSE implementation process. The training also stressed that changing the drought response paradigm in Mexico has involved creating different instruments such as the Natural Disaster Fund (FONDEN) and its equivalent for the agricultural sector (CADENA), as examples of proactive and timely approaches used by PRONACOSE, together with a set of general principles and administrative coordination mechanisms that provide innovative experience that may be useful as models for global efforts to address drought.

These guiding principles are considered the backbone of the policy response to both theoretical and pragmatic considerations as well as of the process to create the main elements of a drought policy, which also can clearly be applied in the Brazilian case and that of other countries facing droughts. These principles include taking a preventive approach, decentralization, governance, training and research, gradualism and evaluation, and institutional coordination. The Mexican process involves two major elements developed according to these principles. They address current drought situations and the transition from certain reactive institutions and rules to mechanisms designed for the new paradigm: (1) elaboration and implementation of PMPMS in every river basin and water utility in the country and (2) mitigation measures to face ongoing drought emergencies.

The program's vision entails planning and implementation of drought measures, involving public participation in the definition of actions to reduce vulnerability as a pillar of the Mexican strategy for adaptation to climate change, as expressed in the General Climate Change Law and the National Water Law and linked with the activities of the National Civil Protection Service. The Mexican program envisions that every basin will have a PMPMS together with periodic evaluation and updates including basin council members' participation to improve the programs and interinstitutional coordination instruments at the national level to prevent and oversee contingencies.

The policy implementation strategy consists of gradually decentralizing attention to droughts by involving stakeholders through the basin councils and pairing them with teams of academic experts from local universities. The goal is to develop local institutional capacity and to begin to change the old reactive top-down approach to address droughts.

PMPMS are the main instrument used to carry out the strategy, and, once they are approved and implemented by the basin councils, the aim is that these councils will move to delineate more specific programs for cities (water utilities) and irrigation districts within that basin. The PMPMS also need to be evaluated and improved within a specific time horizon to assure that the revised programs will be ready before the current federal administration ends and thus be able to continue regardless of administrative changes. While the PMPMS are in place and enough institutional capacity exists at the local level, the interministerial commission at the federal level is in charge of coordinating the federal government mitigation activities.

Finally, the World Bank enlisted NDMC's involvement in the activities in Brazil from the very beginning. NDMC assisted the World Bank in training the Brazilian team and elaborating the initial documents to guide the project. Once its activities were underway, NDMC's role in the learning exchange focused mainly on supporting the creation of the Northeast Drought Monitor. NDMC helped to develop a questionnaire designed to assess stakeholder needs and capabilities, which initiated the process leading toward the eventual development of the Monitor itself. NDMC also participated in a series of webinars aimed at training and preparing the Brazilian team for the Monitor's development. In January 2014, NDMC's Mark Svoboda attended a workshop in Fortaleza with participants from around Northeast Brazil to introduce them to the Monitor concept.

An important component in NDMC's interactions with the Brazilian team was a week-long training that it hosted in Lincoln, Nebraska, in March 2014. Results of this training included: (1) identifying alternatives as to how data and information flow in the production and validation process could be done in Brazil, (2) identifying the relevance and scale required for the developers and validators, (3) identifying the primary indicators and data availability required to calculate indices for Brazil, (4) identifying how to manage the flow of data to translate it into useful products and information, (5) understanding the process of GIS file generation for each of the indicators, and (6) understanding the importance and necessity of local expertise and the incorporation of their information into the Drought Monitor process with a particular emphasis on impact assessment and its role in validating the Monitor outputs. Considerable discussion about the information technology infrastructure and data exchanges required occurred as well.

NDMC also provided advice on the development of drought preparedness plans for various river basins in Northeast Brazil. A main component of these discussions included how to incorporate drought monitoring information, such as that from the Monitor, into the drought preparedness planning process. NDMC furnished a number of vulnerability assessment case studies and stressed the importance of drought impact data collection. Advice on stakeholder engagement and educational activities was likewise provided, as were case studies on the involvement of stakeholders in the drought preparedness planning process.

During the Northeast Drought Monitor development process, NDMC staff participated in multiple webinars providing advice and insights based on USDM experiences. NDMC staff also made several trips to Brazil to provide direct assistance with the Monitor author and validator teams. Mark Svoboda participated in a week-long training in Fortaleza in August 2014. This was the first time potential Monitor authors were brought together from multiple states to receive training on the Monitor process being developed. In addition, NDMC staff provided technical assistance from Lincoln for several workshops in Brazil using webinars both prior to the workshops and connecting with the Brazilian participants at the end of each day, in the process bringing extensive hands-on experience to these meetings from the US perspective. Overall, NDMC staff spent considerable time throughout their interactions with the Brazilian specialists emphasizing the importance of the Drought Monitor process and the key technical, institutional, legal, political, and financial lessons learned from the US experience, together with best practices and information regarding the challenges that needed to be overcome. This support was very important for successful development of the Northeast Drought Monitor, which has moved rapidly toward operationalization.

Finally, NDMC has contributed to the Brazilian efforts by making numerous presentations at international events in recent years. These have included, among others, the Global Water for Food Conference in Seattle, Washington, in October 2014; the American Geophysical Union annual meeting in San Francisco, California, in December 2014, and the European Geophysical Union annual meeting in Vienna, Austria, in April 2015. NDMC's director also highlighted the Brazilian case study during the Global Drought Information System meeting in Pasadena, California, in December 2014. These events have brought favorable attention to the Brazilian efforts.

7.4 Lessons and Recommendations from These Experiences

The lessons learned by Mexico, Spain, and the United States in their drought management efforts over time were invaluable in helping to guide Brazil through its recent drought risk management activities. The principal lessons are briefly described in this section.

Spain, Mexico, and the United States have a long history of dealing with droughts and with adaptation to drought impacts. More recently, however, specific efforts toward a drought risk management approach have been needed in these three countries in order to address and reduce the likely increasingly serious nature of future drought impacts.

Because droughts are complex phenomena and their impacts evolve as societal vulnerabilities change over time in each region, improving drought risk management requires a long-term commitment. As a consequence, short-term projects are likely to meet with limited success. Proactive risk management strategies require years to develop, evolve, be assessed, and be maintained in order to ensure success in reducing drought impacts. This lesson applies both for entities within Brazil, as well as for funding entities such as the World Bank.

Long-term drought risk management can be difficult to develop and advance when the pressure caused by a current severe drought heightens the focus on immediate crisis response. However, it is often a severe drought crisis that provides the impetus to implement longer-term drought risk management.

Likewise, because drought risk management involves a long-term commitment, it is acceptable for initial efforts to be comparatively simple with the expectation that they will build and evolve over time. A prime illustration of this principle is illustrated by the USDM, which was a much simpler process when it started in 1999. It is recommended that the Drought Monitor in Brazil adopt a similar approach as it evolves from being the drought early warning and assessment tool for the Northeast region to potentially cover the entire nation, since, as shown in Chapters 1 and 2, serious droughts can affect other parts of the country as well.

The iterative process that occurs between improved drought monitoring and improved drought risk management through ongoing planning has been demonstrated both in the three nations whose experiences are briefly summarized in this chapter and in Brazil. In robust systems, better drought monitoring naturally leads to better planning as the improved information enters the decision-making process. Likewise, as better drought planning occurs, momentum is created requiring the need for improved drought monitoring and derivative "value added" products as decision makers become familiar with what information can be made available. This also helps to identify areas where greater data and information are needed.

In Mexico, development of the national drought policy provided the momentum to break inertial attitudes both within the river basin councils and in the national government, illustrating the importance of having "top-down" support for drought risk management efforts.

Active stakeholder engagement and transparency related to drought risk management activities help to ensure their success. Efforts to publicize activities, build public awareness and consensus, provide updated tools and information, and develop educational programs were well rewarded when implementing drought risk management.

Finally, continuous engagement with international experience on drought risk management through workshops, conferences, and other fora was extremely beneficial both in building more dynamic drought risk management approaches and infusing updated research and knowledge into these approaches in Brazil and elsewhere.

7.5 Conclusions

Brazil can now use the lessons it has learned from Spain, Mexico, and the United States to share its own lessons of experience with similar countries, regions, and river basins around the world. This is important as organizations such as the World Bank and the IDMP promote drought risk management activities in other locations. Likewise, the World Bank should consider funding more technical cooperation projects like the one in Brazil. Building long-term resilience through better drought monitoring and early warning, tied to improved drought planning and management, is an excellent investment that will provide many quantitative and qualitative returns in the future.

8

Planning for the Next Drought and Paving the Path for Climate Change Resilience

Nathan L. Engle, Erwin De Nys, Antônio Rocha Magalhães, and John Redwood III

CONTENTS

8.1 The Opportunity 91
8.2 Adding Climate Change to the Mix 92
8.3 How Best to Approach the Challenge? 94

8.1 The Opportunity

Recently, prolonged droughts across Brazil have spurred a familiar dialogue within the country to improve drought policy and management. In the past, this conversation has waxed and waned in accordance with the drought cycle, with only incremental progress being made over time to foster more proactive drought management. As has often been the case in the past, the current drought has drawn the attention of the broader Brazilian society, the media, public servants, politicians, and international experts. The key challenge that Brazil, or any country in a similar situation, must face is to find a way to avoid missing this window of opportunity to take bold and progressive action on reforming drought management and planning.

The story of Brazil over the past three years, as told thus far in Chapters 1 through 7 and elaborated upon in the technical section of the book (Chapters 10 through 12), is evidence that the country has taken advantage of this opportunity to show political will and make concrete and lasting advances in the way it approaches droughts. However, significant work remains to be done in Brazil, particularly for the Northeast region, to ensure that recent progress is well engrained into the collective Brazilian psyche, and most importantly, that it is institutionalized into national political processes and socioeconomic dynamics.

The incentives for Brazil to fully achieve this transition to proactive drought policy and management, moreover, extend beyond drought preparedness. It is becoming increasingly evident that shifting the drought paradigm can not only improve the preparation for and response to drought events, but also that having these mechanisms in place will build resilience to a wider range of climate stresses that are likely to become more severe in the decades ahead. As with the phenomenon of drought, climate change manifests itself over longer timescales, is difficult to define with respect to impact attribution, and is a "creeping" phenomenon (i.e., it is not well-detected until it is advanced and widespread) that can catch decision makers and stakeholders off guard. The manner in which a nation, community, or individual decision maker approaches droughts through governance, institutions, policies, investments, and strategic choices is a harbinger for how a society might approach the problem of climate change. Therefore, droughts and the ways in which they are managed can be seen as adaptation catalysts for laying the building blocks of improved climate change management (Engle 2012).

8.2 Adding Climate Change to the Mix

The technical networks, institutional channels, and operational flow of information and resources that are improved through the three pillars of drought preparedness are similar to what is needed more broadly for managing other climate risks. It is well understood that effective institutions and governance play an important role in reducing vulnerabilities to climate change and extreme droughts, as well as for sustainable development more generally (World Bank 2003, 2010). Importantly, there is a need for strong, inclusive, and collaborative institutions that are capable of planning and implementing a range of possible technical, economic, and legal approaches, in order to respond to growing water scarcity, bring about greater drought preparedness, and build climate resilience. Two examples of this are the hydrometeorological and agricultural early warning and information systems/networks built via the Northeast Drought Monitor, and the interinstitutional/governance mechanisms for managing droughts fostered via the national/regional dialogue and various participatory processes associated with drought preparedness plans. Securing both of these advances will pay dividends for managing other climate risks.

But just how critical is the issue of climate change to Brazil? The World Bank's third and most recent publication that analyses climate change impacts associated with 2°C and 4°C future warming, the "Turn Down the Heat Report," indicates that climate change is projected to significantly impact the Latin America and Caribbean regions. For the most part, dry regions will become drier and wet regions wetter, except for central Brazil (where annual mean precipitation could decrease by 20% in a 4°C world by 2100). Also by 2100, the most of Brazil (other than its southern coast), southern Chile, the Caribbean, Central America, and northern Mexico will likely experience severe to extreme drought conditions relative to the present climate, and arid and semiarid lands could expand 8% in a 4°C world (World Bank 2014). This is in line with the results of similar studies that show the Brazilian northeast experiencing reductions in mean annual precipitation combined with increased mean annual evapotranspiration, ultimately suggesting an increased likelihood of droughts over the coming decades (World Bank 2013). Higher drought risks will also increase the likelihood and severity of forest fires, forest degradation, and ecosystem service losses. Combined, the increasing droughts and extreme temperatures are projected to lead to more cattle deaths, crop yield declines, and water availability stress and insecurity with serious associated adverse economic and social effects.

With respect to agriculture specifically, recent World Bank data show that the negative impacts on production are disproportionately large for Northeast Brazil (compared to the rest of Brazil), mainly due to a projected overall reduced rainfall and shifts in precipitation patterns and micro dry periods during existing cropping cycles. Even 2°C warming could reduce crop yields by 50% for wheat and 70% for soybeans (World Bank 2014). One World Bank study in particular explores simulated adaptation strategies for agriculture in the Northeast region, and shows that the use of supplementary irrigation, soil and water management, and changed cropping cycles could reduce the projected negative impacts of climate change out to 2050 (Fernandes et al. 2012). These findings highlight the nuanced nature of projected climate change impacts within the Northeast region, with some areas expected to be more negatively impacted than others. Not only does this imply that understanding these spatial patterns will be key to more cost-effective adaptation strategies, but also that some areas in the region will likely not be viable at all for agriculture in the coming decades. This has implications for how to target resilience-building efforts in these communities, particularly in the context of increasing droughts. The processes involved with designing and implementing context-specific drought preparedness measures, like those discussed in this book, will help decision makers navigate this process more effectively.

Important challenges still remain regarding how to develop planning and management strategies for longer-term climate resilience and on how to harmonize such processes and priorities to ensure sustainable water allocation decisions are also considered in the context of drought and water shortages, energy supply and demand, ecosystem needs, and regional economic development policies and programs. At present, the discrepancy between the design assumptions of water allocation planning and the operational reality of needing to manage greater uncertainty with climate change impacts and demand increases, for example, leaves little margin of maneuver and is often a main driver of water conflict. Analyzing, documenting, and understanding the key vulnerabilities across sectors and projects will help facilitate adaptation to the hydrological effects of climate change, particularly to increasing droughts and uncertainty of future water supplies.

8.3 How Best to Approach the Challenge?

Still, whether it is better to achieve this harmonization through separate but coordinated planning processes, or through one combined strategy, has yet to be determined. Several different options are available for Brazil to further institutionalize drought preparedness and integrate these processes and priorities among and between the numerous drought-related sectors of Brazil's society and economy.

Although issues related to climate change are now firmly on the political agenda in Brazil, efforts to date have concentrated mainly on responding to short-term climate variability (principally in terms of drought responses). Climate change is not yet fully factored into adaptation action plans or long-term decisions across sectors. Brazil is, however, in the process of completing its first ever national climate change adaptation plan, a process that would be well suited to incorporate the elements of drought preparedness.

Drought policies and preparedness planning involves mitigating drought risk through improved response and relief, but also through longer-term resilience building and adaptation measures. As noted earlier in Section 8.1, drought preparedness in itself can also contribute to broader climate change resilience building within a country. This suggests the need for a debate within Brazil and elsewhere on which structural changes will best serve to mitigate future drought and other climate risks, and how these measures should be prioritized to increase broader resilience through a climate change adaptation plan, a comprehensive drought plan, or separate drought plans on a sector-by-sector basis (or some combination of all three). In the water sector, for example, decision makers could debate the applicability and relationship of a national drought policy in the context of efforts to define water management strategies for addressing long-term drought and climate uncertainty. This could include the elaboration of the National Water Security Plan, the operation of large-scale interbasin transfer projects like the São Francisco Inter Basin Water Transfer Project, and the revision of the National Water Law and National Water Resources Management System. In any case, improving institutional capacity for planning and coordination of short- and long-term response actions is key to the implementation of a proactive drought policy.

Future droughts are likely to bring greater stress on water resources as extreme climate variability and climate change collide with increasing water demand from population growth and regional economic development. Now, more than ever, there is a need for Brazilian officials and society to continue down the path started during the past three years to rigorously discuss how to institutionalize and integrate comprehensive policy on drought planning and management in order to proactively increase resilience to future droughts and climate change.

The drought currently affecting the Northeast region is the most intense in decades. Serious drought and associated water shortage conditions have also plagued other parts of Brazil in recent years. Along with related federal and state responses to droughts, the increased attention worldwide to draft and implement coordinated national drought policies that embrace the concept of drought preparedness, national and international attention on building climate change resilience, and the country's recent technical and institutional advances, Brazil now has the necessary ingredients to better integrate proactive approaches to drought risk management. Until recently, the right combination of these ingredients had eluded policy officials, public servants, and the broader Brazilian society in attempts to move ahead on drought policy and management. The convergence of these efforts, interests, and capacities presents a unique opportunity for Brazil to make significant progress in the years ahead.

9

Voices of the People: Socioeconomic Implications of Drought in Northeast Brazil

Dorte Verner

CONTENTS

9.1 Water Scarcity 100
9.2 Economic and Livelihood Implications 102
9.3 Health Implications 104
9.4 Social Protection 106
9.5 Migration 106

Drought affects most people's livelihoods and well being in the semiarid *sertão* of Northeast Brazil. Multiyear droughts often wipe out livelihoods, deplete financial and other assets, and harm human and animal health. In worst cases, it leads to death. A few images of the current multiyear drought from my trips to Ceará are shown at the end of this chapter (Figures 9.1 through 9.20). These images illustrate drought impacts such as reduced to no yield in agriculture and other livelihood effects; water shortages; and time use, labor, migration, health, and other socioeconomic implications.

For the poor particularly, the detrimental effect of climate change on the environment erodes a broad set of assets—natural, physical, financial, human, social, and cultural. With climate change, including drought, coping strategies that people have adopted throughout history may no longer be available and adequate; therefore, climate adaptation is needed. Brazil's long history of interventions are nevertheless showing results as the current and severe multiyear drought is pushing fewer people to migrate out of the Northeast drylands.

Droughts are nothing new to the people in the *sertão*. They have been recorded at least since the sixteenth century, and the resilience of poor *sertanejos* (people of the *sertão*) to climate shocks is relatively low (see Figure 9.2). Many live in environmentally fragile areas that are especially prone to natural hazards such as droughts and floods, and many depend directly on fragile natural resources for their livelihoods and well being. When circumstances change for the worse, the poor are often hard put to adapt. For many, the effects of droughts are compounded by other pressures. Those interviewed for this chapter mentioned the increasing scarcity of viable land for agriculture, joblessness, poor health, limited education and skills attainment, social marginalization, and lack of access to credit and insurance as significant constraints.

A drought of the severity of the current one has not been experienced in the Northeast states for many decades, and possibly for over a century. Recurrent droughts have a huge impact on rural people's lives and livelihoods. Everywhere, poor people are disproportionally affected in the absence of good development policies and social safety nets. Although throughout history livelihoods have adapted to change, without appropriate public sector interventions, it is very likely that the future impacts of climate change will push poor people beyond their capacity to cope (see Boxes 9.1 and 9.2).

BOX 9.1 NORTHEAST MUNICIPALITIES WILL BE HARDER HIT BY CLIMATE CHANGE THAN MUNICIPALITIES IN SOUTHERN AND SOUTHEASTERN BRAZIL

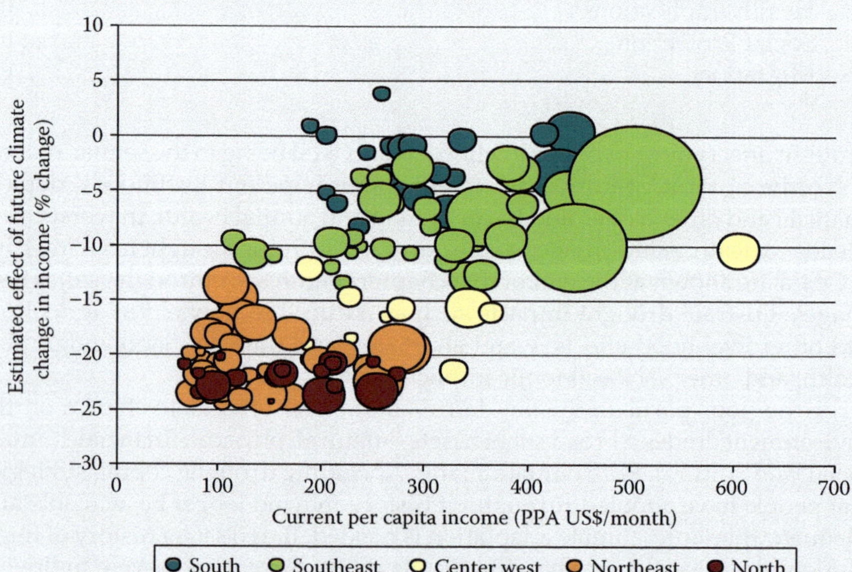

Economic implications of drought and other forms of climate change are likely to be huge in the Northeast states. In Brazil, the average impact of climate change on per capita income is estimated to be a 12% less by 2058. Per capita incomes of poor municipalities are likely to be more adversely affected by climate change than those of richer municipalities. The municipalities in the Northeast and North of Brazil will likely be the worst hit. Per capita incomes in these municipalities could be reduced by up to 23% from their already low levels. This contrasts with

(Continued)

the situation in the larger and richer municipalities in the Southeast regions that are likely to experience much smaller adverse impacts and would have an easier time adapting to these adverse effects on their economy. In the absence of appropriate adaptation actions, climate change will, therefore, likely contribute to increased regional disparities. Finally, a comparison between urban and rural farm and nonfarm households shows that the rural households are likely to be the hardest hit by climate change.

Sources: Verner (2010) and Andersen and Verner (2010).

BOX 9.2 CLIMATE CHANGE WILL LIKELY IMPLY HUGE ECONOMIC COSTS FOR HOUSEHOLDS AND THE OVERALL ECONOMY

Andersen and Verner (2010) analyzed past temperature and precipitation trends for 50 years for 34 high-quality meteorological stations in Brazil and found that of these, 31 stations show significant warming, 3 show no significant change, and none show significant cooling. The authors found that the North is warming about twice as fast as the South and the Northeast and Center West regions are warming at intermediate rates. In contrast to the results for temperature, the authors found no clear tendencies with respect to precipitation.

The climate change impacts on agricultural production (yield changes), and international food prices will be large unless proper mitigation measures are implemented. The specific impacts of climate change on Latin American economies, agriculture, and people were analyzed by Andersen et al. (2014). Findings show that Brazil may face economic losses of US$272.7–550.6 billion by 2050. At the household level, Brazilian households may lose 4.3%–28.8% of their annual incomes because of climate change by 2050.

The gender analysis suggests that male-headed households may be more vulnerable (less resilient) to climate change than female-headed households, as female-headed households tend to have slightly higher per capita incomes and higher levels of income diversification than their male counterparts in Brazil.

Sources: Andersen et al. (2010, 2015).

Although drought policy and management in Brazil need to be more proactive and less fragmented, northeasterners' resilience has increased in recent decades compared with earlier droughts and migration from the *sertão*. As highlighted throughout this book, climate change calls for strong leadership to build resilient households and communities, and the Brazilian government's interventions are working. The scenario of the past centuries when thousands of people were forced to migrate is only partly seen today, as poverty reduction and other development programs have put systems in place that have curtailed migration from the *sertão*. This chapter is based on visits to the region and communication with people of all ages during the current extended drought that is now moving into its 6th year. The write-up is mainly based on conversations and stories by people about their reality in the current prolonged drought. These personal stories are often told with comparisons to earlier droughts (see Chapter 1 for the list of drought years in recent times). This chapter addresses five areas related to the resilience of people during droughts: water, livelihoods, social protection, health, and migration.

9.1 Water Scarcity

Water scarcity is traditionally a critical issue during multiyear droughts for families in the *sertão*. This is also the case for the current drought. Water availability is generally reduced progressively for each year of drought. Water is essential for the survival of both humans and domestic and wild animals. The lack of, or sheer absence of access to, water has pushed thousands of families to migrate during the past centuries (see Section 9.5). The most common past and traditional measures taken in an effort to alleviate this situation are the financing of tank trucks (*carros pipas*) to distribute water and the drilling of additional wells.

As was described in Chapter 1, over the past few decades, road and water infrastructure investments have increased and dams and reservoirs have become widespread throughout the region. As a result, water can now be more easily distributed to households than during previous droughts. During the current drought, water is still a serious issue for most people; however, during the initial years of the present drought, plenty of water was still available in the dams and reservoirs for distribution to people in the region. But with each passing year with little rainfall, the severity of the drought is increasing. People's water needs do not fall during droughts. In contrast, they increase, and drought impacts are not only a water scarcity issue but a life and livelihood issue as well.

With the above mentioned implementation of road and water resource infrastructure, water can be distributed much more easily than in the previous decades. As mentioned in Chapter 1, government programs have been put in place to supply water to households and municipality centers. The *carros pipas* can be seen everywhere during drought periods, both public and private ones. These trucks come every two days to many places and fill up home cisterns used for water storage. This provides families with better access to water and helps alleviate their difficulties to get through a drought period. These cisterns are often located in the municipal seats (see Figures 9.3, 9.4 and 9.7), as well as on private properties in the rural areas (see Figure 9.8).

In recent years, the federal government, through the Ministry of National Integration, has been supplying cisterns to households for water storage for both human and animal uses. As this book is being written, there are many households that are digging in the hard ground in order to have a cistern installed. The quality of the water, however, is sometimes mentioned by users as being inferior, and households are complaining. Some people called it muddy water and explained during the conversations recorded for this chapter that it likely comes from lakes or other places without treatment. Some mentioned that even when the households treat the water, it is undrinkable. Having a cistern installed on the property nevertheless makes a huge difference compared with not having one (see Figure 9.8).

As the drought progresses, however, it becomes increasingly difficult to obtain water sufficient even for human use in most places. Those who do not have a cistern frequently have to travel to obtain water, for example, from the cistern at the center of the nearest town. Or they have to buy water, which is expensive. Some of those interviewed mentioned that 75% of the *Bolsa Família* stipend was used to pay for water (see Section 9.4).

In many rural households, people—often women and children—have to travel far to obtain water for their family and domestic animals (see Figure 9.5). Traveling to collect water is time consuming for many women and men. Also, children are helping their families, which often means that they miss school. Throughout the region, women can be seen carrying large buckets of water on their heads or pushing wheelbarrows containing multiple buckets. Often men carry two buckets of water using a pole or shoulder yoke, with one bucket on each side (see Figure 9.3). Occasionally men use donkeys to carry water buckets. This picture has changed little over the past decades except that there are fewer people doing this presently as more households now own cisterns.

Water is scarce for human and animal consumption and for washing clothes (see Figure 9.6). During the current drought, many women are taking their dirty laundry to places where there is water, including where it is being pumped from a well. Although this also happened two to three decades ago, the amount of clothes that families now own has increased as households receive more income (see Section 9.4).

9.2 Economic and Livelihood Implications

Most people's livelihoods and incomes are affected when a drought occurs, including those directly and indirectly involved in sectors that are highly dependent on natural resources, such as agriculture and livestock. Agriculture started in the areas that today are similar to Northeast Brazil, namely the Middle East, 8500 years ago. A few thousand years ago, a 300–400 year-long dry cycle likely led to a drastic change away from agriculture to pastoralism. Livestock are more resilient than crops during short droughts, but, in multiyear droughts, they too is vulnerable. Over time, the population and livestock herds grew and borders were drawn up limiting the space where the herders could move freely. When droughts hit, carrying capacity of the land is severely challenged. Today most Arab nations face drought on a yearly or bi-yearly basis in parts of the countries. The Northeast states are not yet there (see Figure 9.9). Moreover, Brazil has a long history of taking action aimed at reducing drought impacts. Traditional measures to help address the drought situation in the Northeast states include opening emergency credit lines and renegotiating farmers' debts; measures that are noted throughout this book.

Drought puts additional pressure on scarce resources in the semiarid Northeast states. During prolonged droughts, small farmers lose their planted crops and experience extremely low yields, if any. Their families and other farm worker families often experience a dire food security situation and need urgent food assistance. Prolonged drought also forces pastoralists to sell their animals, often at extremely low prices, run down other assets buying feed, or see their livestock die because of lack of pastures. Soaring feed costs and food price shocks worsen the situation. Household incomes generally plummet in these extreme drought periods. In addition, droughts cause health shocks that affect labor productivity and other types of well being. Therefore, people's vulnerability increases. Climate, in fact, is involved in many of the shocks that drive or keep people in poverty (World Bank 2015c).

Experience with droughts from across the globe illustrates the importance of increasing livelihood resilience when confronted with climate-related disasters. As Chapter 8 shows, this situation emphasizes the need to foster adaptation as well as to develop livelihood resilience. Despite the adaptation policies that are in place in the Northeast states, many communities remain vulnerable to drought. Droughts have huge livelihood implications because people rely heavily on natural capital; some 40% of the economically active population in the region depends on agriculture (Lemos 2007). In Ceará, which has been experiencing a drought on average every 3½ years, as much as 96% of the agriculture is rain-fed, and an estimated 90% of small-scale farmers (many of whom are tenants) have no source of employment-related income outside their farms (Brant 2007). This strong dependence on natural capital, together with the great risks associated with climate variability (i.e., variations in the timing, location, and amount of precipitation), leaves many families vulnerable to livelihood shocks. More than half of the households in the Northeast states are considered food insecure (IBGE 2004). In this region, which is dominated by subsistence farming practices, long-term drought has significant livelihood implications and exacerbates existing tendencies such as lack of long-term planning and investments, neglect or depletion of human capital, dependence on state transfers, and migration as a means of survival.

Droughts are a shock to the local economy as many jobs disappear and farmers experience lower productivity or even lose entire harvests. Increasing heat and drought threaten the livelihoods of already marginal smallholders. This retards the agricultural growth, which impacts other sectors as the purchasing power of those in the agricultural sector is reduced. These costs and impacts are quantified for the current drought in Northeast Brazil in Chapter 10.

With severely reduced water resources and rainfall, drought has a profound effect on livelihoods, and subsistence farmers see their livelihoods wither as their crops are rain-fed. Shortages of water and animal fodder are limiting factors for small farmers during droughts. In many countries, there are limited policies and often they are restricted to small amounts of animal fodder and water. Even in Brazil, government initiatives are not sufficient to meet the needs of rural people during droughts, although progress has been made (see more on programs in this chapter). Farmers and rural workers have become used to depleting their limited assets. For centuries, when assets were depleted, the last resort for most families was to migrate from the *sertão* to urban areas.

Farmers in the *sertão* mainly plant and harvest corn, beans, and cassava (see Figures 9.10 and 9.12 through 9.14). During droughts, this becomes increasingly challenging as the question of when to plant, if at all, becomes a difficult decision. Previously, farmers based their planting timing decision on very little information apart from tradition. Today, information technologies allow meteorological information to flow fast, and most farmers can obtain weather forecasts and other information on a daily basis. During droughts, farmers also use this information and their own experience to decide when to plant. Most farmers in rain-fed areas want to plant. Many farmers therefore do plant during the drought. Some 1.1 million farmers in the Northeast states benefit from *Garantia Safra*, a crop insurance scheme (see Chapter 2).

Farmers interviewed for this chapter mentioned that in some drought years they plant and get insurance, whereas in other drought years they also plant but do not get insurance. People in the Northeast states know droughts and their impacts as all have lived and experienced drought impacts. Many people mentioned that the droughts in 1932 and 1958 were the worst ones they had heard about or could remember as nothing was harvested in those years. Some farmers also recalled that, in the early 1980s, there was so much rain that crops rotted in the field.

During serious drought periods, pastoralists have lost all their livestock. Livestock suffers as farmers cannot afford to buy all the food and water needed for the animals. Often they die because of undernourishment. In some places up to 70% of the herd died in recent droughts. Based on information provided by farmers during the current drought, farmers try to sell their animals before the price gets to the point of no return, or buy fodder, cut grass, and try to find other ways to feed their animals (see Figure 9.11). Livestock have always been important for households as they have served as a form of reserve capital or bank account in which savings are stored as livestock, which are more resilient to shorter droughts than crops.

Fruit has always been an important part of the diet in the Northeast states. Trees bear little or no fruit in drought periods. For example, mango trees bear fruit only when it rains in the winter, so in the past years the trees have hardly borne any fruit. This has implications for the nutritional status of the people who eat fewer fruit in drought years than in nondrought years.

9.3 Health Implications

Life in the *sertão* used to be extremely precarious because of the prevailing hunger, thirst, and diarrhea, but fortunately this has mostly changed (see Figures 9.15 and 9.16). Hunger has nevertheless been experienced by most adults, especially by those over 40. A number of adults interviewed mentioned that at times when they were children, they could not attend school because they were too weak to walk due to hunger.

Droughts have serious health implications related to the fact that the water resources are scarce and often can be of lower quality, as mentioned earlier. Many *sertanejos,* who collect water from reservoirs, do not treat it before drinking. Some use water filters that improve water quality for drinking. If the water is not properly treated, children's health suffers, for example from diarrhea. One of the leading causes of child mortality in the *sertão* used to be dehydration because of diarrhea.

Infant mortality has been a very important issue. Families used to be much larger than they are today, and some families lost a large number of children. One 64-year-old man recalled that he had lost three of his nine children. There used to be a child death per day according to one of the *Agentes de Saúde* (community health agents) interviewed (see Figure 9.18).

Child and infant mortality in the 1970s was among the highest in the world in parts of the *sertão.* Some municipalities had an infant mortality rate (IMR) of more than 400, that is, 40% of all infants died in the first 12 months of life. Since then, the IMR has fallen faster in the Northeast states than in any other part of Brazil; from 112 in a thousand live-born infants in 1990 to about 20 in 2010 (although this figure is still above the national average). Families often did not give their newborn a name (Scheper-Hughes 1992). People recall that some parents did not want to spend the money to register the child before it got older. When infants died, they often were buried in a small coffin lined with blue material symbolizing that the *anjinho* (little angel) was going to rise to heaven. The *Viva Criança* (Long Live Children) program across Ceará made a huge difference in reducing IMRs. Low incomes and education levels played a role, as families and mothers usually did not have what they needed to save their children's lives. They were not aware that an oral rehydration solution made of salt, sugar, and water could save their children from dying of dehydration because of diarrhea. Community *Agentes de Saúde* (health agents) now go from house to house in villages throughout Ceará and teach mothers health skills. The agent records births, deaths, and general health problems and sends the sick to the nearest health center or hospital as needed. During drought years, community health agents put a lot of emphasis on the filtration of water for the rehydration mix and other measures to ensure that infants do not suffer.

The lack of food was another main cause of the high IMR. Now, with programs such as *Agentes de Saúde* and *Programa de Leite* (the milk program, through which a liter of goat milk is distributed per child per day), much more attention is being given to early childhood health needs, contributing to the survival of some of the most vulnerable. This, together with increased household incomes through social protection programs (see Section 9.4) and community health agents, has contributed to a huge drop in IMRs and better health overall in Northeast Brazil.

The fertility rate has also plummeted in the Northeast states. For example, now most women have only one to three children, and these children usually all survive. After the third child, most women get sterilized as a way to prevent having more children. This compares with their mothers, who often had about 10 pregnancies, 8 live births, and just 4–5 babies who lived beyond their first year.

Until recently, houses were frequently made of *taipa* (a combination of straw, mud, and sticks) and had no water, electricity, or sanitation, and families had little food security, especially during drought periods (see Figure 9.17). When farmers have a good harvest, they sell beans and maize and buy other needed items such as fruit and nutritious food. Social programs cannot make up for all these losses because of the drought. This means that families cut down on buying fruit and other food items. Their diets suffer during drought periods as most families do not have enough financial resources to buy what is needed to sustain the same diet as in a drought-free year.

Diseases such as Chagas, cholera, and tuberculosis used to be common in Northeast Brazil. Now they are rare, but during drought periods waterborne diseases such as cholera can occasionally occur.

9.4 Social Protection

During drought periods, there is an increase in social transfers as the vulnerability increases in drought-stricken communities in the northeastern states. These are in addition to the social protection programs that are part of the overall human capital building and poverty reduction efforts in Brazil.

Children used to be an essential part of the labor force in the Northeast states. Over the past two decades, however, child labor has virtually disappeared. The *Peti* (Programa de Erradicação do Trabalho Infantil; Child Labor Eradication Program) was one of the first programs to combat child labor. It drew attention to the issue and provided families with a stipend if children stayed at school full-time. The *Bolsa Família* (family grant; the family allowance income transfer program), which provides conditional cash transfers to low-income families as long as they vaccinate their children and retain them in school, also provides cash to drought-affected households with children.

In addition, the rural pensions program has provided income to people who have traditionally had very little cash on hand. Women and men who have worked in agriculture are entitled to a government pension when they reach 55 and 60 years of age, respectively. As a result, the vast majority of these farmers, who have had very little income and thus very little to spend, now receive money every month. Given their previous low propensity to consume and considering that most of their limited financial resources have been used for agricultural production, this added income has frequently led to increased spending for children and grandchildren and clothes are often the main focus of the increased consumption.

9.5 Migration

Climate-driven migration from the *sertão* has changed signficantly over time, from being largely a response to distress to more of a search for off-farm employment, and has tended to decline in recent decades due to decreasing demographic pressures (see Figures 9.19 and 9.20). Thus, climate-driven migration can, for simplification, be divided into two broad categories: labor migration and distress migration that may be either temporary or permanent. Climate-induced labor migration is an adaptive response, typically taken as a last resort, by households confronted by climate-related stress such as gradual or chronic drought. Climate-induced distress migration is a result of natural disasters. These migrant flows comprise large numbers of distressed people seeking aid until they may be able to return to their homes. The characteristics of distress migration differ over time, as they are shaped by the severity of a crisis, the ability of individual households to respond, opportunities, existing and persisting vulnerabilities, available safety nets, and intervening government policies.

For centuries, drought-related migration from the *sertão* could be characterized as distress migration and was undertaken by foot. It was extremely arduous in the hot weather with little availability of water and food and many people lost a great deal of their physical strength because of poor nutrition. Therefore, many people recall what a struggle it was to get to the cities on the coast or to southern Brazil. Many people died along the way.

Today most drought-related migration takes the form of temporary labor migration. Some young people leave their homes in the Northeast states to work in agriculture harvesting, such as cutting sugarcane in the Southeast states or picking oranges in the center west of Brazil, and send money back home to their families. These seasonal migration patterns are different from earlier decades when men often left and sent money home, so that their families could eat better. Then later they were followed by their whole families migrating to them. They left for good with their few belongings trucks or buses going to the coastal cities or cities in the South.

Migration can provide several benefits for migrants and their families. A large share of households in drought-stricken areas receive income from relatives outside the region; for example, 31% of all households in Northeast Brazil frequently receive some form of remittances (Lemos 2007). Remittances to the family left at home will raise the household's income, perhaps improving food security, access to health care, or allowing children to go to school instead of working. Physical assets such as housing quality or livestock may also be improved or increased (Verner 2010).

The experience in Northeast Brazil illustrates the importance of adaptation in order to increase resilience when confronted with climate-related hazards. The drought situation also emphasizes the importance of social programs as part of a regional climate change adaptation strategy in order to enhance coping capacity of households. The rural pensions program and *Bolsa Família* have also contributed directly to increase drought resilience and reduce migration from drought-stricken areas during the recent prolonged drought period (Verner and Tebaldi 2015).

FIGURE 9.1
The multiyear drought wipes out livelihoods.

FIGURE 9.2
Droughts deplete assets and harm health.

FIGURE 9.3
In rural households, people travel far to get water.

FIGURE 9.4
A municipal seat's water storage that improves water access.

FIGURE 9.5
Women and children collect water.

FIGURE 9.6
Water is scarce also for washing of clothes.

FIGURE 9.7
Cisterns alleviate struggle in a drought period.

FIGURE 9.8
Some private properties also have cisterns.

FIGURE 9.9
Preparation of the land while waiting for the rain.

FIGURE 9.10
In the *sertão*, farmers plant corn, beans and cassava.

FIGURE 9.11
When fodder is limited, farmers cut grass to feed animals.

FIGURE 9.12
Drying of a meager bean harvest during drought.

FIGURE 9.13
Beans are the main source of protein when income is low.

FIGURE 9.14
Beans are rain-fed and yields are limited during droughts.

FIGURE 9.15
A drought-stricken field.

FIGURE 9.16
A little water attracts animals and people alike.

FIGURE 9.17
Family in their house made of straw, mud and sticks.

FIGURE 9.18
Community health agents in a rural settlement.

FIGURE 9.19
People at a truck stop.

FIGURE 9.20
Climate-induced migration has been reduced over time.

10

Drought Impacts and Cost Analysis for Northeast Brazil

Paulo Bastos

CONTENTS

10.1 Introduction 119
10.2 Pertinent Literature 120
10.3 Data Sources 123
10.4 Empirical Methodology 124
10.5 Empirical Results 125
  10.5.1 Trends in Drought Indicators 125
  10.5.2 Trends in the Volume of Water Reservoirs Relative to Capacity 128
  10.5.3 Trends in Main Economic Indicators 129
  10.5.4 Econometric Estimates on the Economic Impacts of Drought 135
10.6 Policy Response to the Current Drought: Main Actions and Associated Cost 139
10.7 Summary and Conclusion 141

10.1 Introduction

One of the most severe multiyear droughts in decades has been plaguing Northeast Brazil since 2010. Studies have been released indicating that the region will likely suffer even more prolonged droughts and water resource stress as a result of climate change (IPCC 2013). Combined with the relatively poorer socioeconomic conditions of the region, these factors leave Northeast Brazil especially vulnerable to extreme droughts.

The region has a long history of managing and living among these challenging conditions, including the introduction of water storage and transfer projects and the advent of innovative institutions and programs to deal with water and climate variability. However, when extreme droughts hit Northeast Brazil, the structural solutions of the past decades, while necessary, are insufficient to overcome these multiyear periods of below average rainfall. Like most other nations, Brazil has approached the management of the multiyear drought events that occur every several decades through various emergency relief and response activities, as summarized in Chapter 1.

The impacts of the droughts and associated emergency response activities are perceived to come at a high cost to society. However, it has been difficult to develop a robust and multisectoral analysis that quantifies these impacts and costs in Northeast Brazil, partially because of the difficulty in determining the onset of the droughts, and also due to methodological challenges in quantifying and attributing impacts to a specific drought. As a result, it has been difficult for society and decision makers to assess the costs and impacts of droughts in the region. This chapter contributes to fill this gap. It combines several sources of detailed official data to assess the economic impacts and costs of the most recent multiyear drought in Northeast Brazil.

This chapter is organized as follows. Section 10.2 reviews the existing literature seeking to quantify the economic impacts of droughts. Section 10.3 describes the data utilized. Section 10.4 outlines the empirical methodology used to assess drought impacts. Section 10.5 presents the empirical results. Section 10.6 describes the main features of the policy response to the most recent drought and seeks to assess the associated budgetary costs. Section 10.7 presents the conclusions.

10.2 Pertinent Literature

There is a growing body of literature devoted to estimating the impacts of droughts on economic outcomes. This literature can be divided in two main types. First, there are descriptive case studies that document losses in indicators such as agricultural production and livestock relative to a reference year characterized by normal precipitation levels. Second, there are econometric studies relating the evolution of geographically disaggregated longitudinal data on precipitation (or drought indicators based on precipitation) to the variation in local economic outcomes.

The first set of studies builds on the assumption that deviations in the observed agricultural output in the years characterized by abnormally low levels of precipitation (relative to a reference year) can be attributed to droughts. Under this assumption, a simple before and after comparison of agricultural outputs gives an estimate of the losses imposed by the drought. On the other hand, studies resorting to econometric methods recognize that other simultaneous, time-varying shocks can influence changes in agricultural output. They also employ more rigorous statistical methods to isolate the causal impact of droughts at the local level. Because application of these methods requires the use of historical series of spatially disaggregated data, the estimates correspond to average local effects of a drought of different degrees of severity.

A number of descriptive studies examine losses in agricultural production associated with droughts in Northeast Brazil. Magalhães and Glantz (1992) provide evidence on the impacts of the 1979–1983 drought, one of the longest ever observed in the region. They find that the drought was associated with sizable losses in agricultural production: between 1978 (normal year) and 1980, production losses amounted to 72% in beans, 82% in corn, 52% in rice, and 70% in cotton. Sarmento (2007) examines the role of drought in explaining the evolution of agricultural value added at the state level over the period 1970–2001. To compute the cumulative loss of agricultural value added that can be reasonably attributed to drought, the study considers only drought years in which agricultural value added experienced a negative growth (relative to the previous year). By applying this criterion to the states of Northeast Brazil (with the exception of Maranhão), the study finds that the cumulative loss in agricultural value added that can be attributed to drought over this period amounted to $13.2 billion. Three states from the semiarid region (Rio Grande do Norte, Paraíba, and Ceará) accounted for 52% of this cumulative loss. In a related study, Khan et al. (2005) examines the impact of the 1998 and 2001 droughts on agricultural production, employment, and income in the microregion of Brejo Santo and in the state of Ceará as a whole. The empirical analysis draws on survey data from official sources and uses the year 2000 (which had normal precipitation levels) as the reference year for computing losses for several crops (rice, beans, corn, and cotton). The results reveal that the 2001 drought led to estimated losses in production revenue of about 70% relative to potential. The losses in physical agricultural output and employment were found to be of similar orders of magnitude, though somewhat heterogeneous across crops. In comparison with the 1998 drought, the 2001 drought was associated with larger losses in agricultural production and revenue, but with similar employment losses.

Ximenes et al. (2013) employ a similar methodology to quantify losses in agricultural output in Northeast Brazil since 1991. Using historical precipitation data since 1960, they first identify drought years based on negative deviations of annual levels of rainfall from the historical mean. Using data from national agricultural surveys, they then quantify losses in agricultural production associated with droughts by computing the relative or absolute deviation of production quantities and values. In doing so, they distinguish between losses in temporary and perennial crops. The results indicate that, relative to 2011, physical output of temporary crops (measured in tons) fell by 13% per year in 2012 and 2013. These losses were clearly larger than in 2010, when output of temporary crops fell by only 1.5% relative to 2009. But they were cumulatively smaller than for the 1992–1993 drought when output losses amounted to 4.6% and 40.1%, respectively (relative to 1991).

Econometric studies in this domain are scarcer. Using data from three independent cross-sectional household surveys conducted in 1992, 1993, and 1995,

Mueller and Osgood (2009) find that past occurrences of drought in a given federal state had persistently negative wage effects. The demographic and income data consisted of three independent cross sections of approximately 330,000 workers living in both rural and urban municipalities of Brazil. The geographic scope of these surveys made it possible to control for regional variation in climate mitigation and adaptation. The econometric analysis relied on the climate variation between cross-sectional surveys from three different years. In particular, the authors created four drought indicators: the average number of standard deviations below the precipitation mean observed in several periods prior to the period of analysis (1–5 years before, 6–10 years before, 11–15 years before, and 16–20 years before). Defining the drought variables in terms of the elapsed time rather than the specific drought years made it possible to identify long-term wage effects of drought duration. The methods employed also accounted for unobserved spatial and temporal heterogeneity by including state and year fixed effects. The econometric results indicated that droughts depressed wages in rural municipalities for 5 years after the event. In addition, they found that dependence on agriculture has had a significant influence on the magnitude of these impacts. It should be noted, however, that the data and econometric approach adopted in this study had some important limitations for estimating the long-term effects of drought on wages. In particular, the use of cross-sectional data on the outcomes of interest made it difficult to isolate the causal effect of drought: if municipalities that experienced drought in the past differed from those that did not in other important respects (such as local institutions and geography), it was not possible to isolate the causal impact of drought from that of other factors.

In a more recent study for Brazil, Bastos et al. (2013) examined long-term effects of droughts on agricultural value added and local labor markets over the period 1970–2010. Using rainfall data going back over a century, they constructed contemporaneous and historical drought indices for more than 3000 local areas and examined them in conjunction with the five most recent sets of population census data. Contemporaneous drought indexes capture the occurrence of droughts in the census years, whereas the historical ones quantify the cumulative incidence of droughts during the previous decade. Based on a differences-in-differences econometric model that accounted for unobserved heterogeneity of local areas along with random trends at the local and regional levels, they found that a higher frequency of droughts during the previous decade significantly reduced local agricultural value added in rural areas.* In addition, employment and wages were estimated

* In a related study for India, Burgess et al. (2011) examine economic and social impacts of precipitation and temperature fluctuations across districts in the period 1957–2000. Using fixed-effect econometric methods that seek to account for the potential influence of factors other than weather variation (such as changes in international agricultural prices, or national and international demand), they provide evidence that dry weather conditions in rural areas significantly depress contemporaneous agricultural output and wages, and consequently raise local agricultural prices.

to fall in the local manufacturing and services sectors, pointing to the existence of negative spillover effects of drought on the whole local economy.* By reducing employment opportunities and lowering local wages, droughts can also be expected to cause migration from the drought-affected areas. Bastos et al. (2013) examined this prediction and found that a higher frequency of droughts in the previous decade led to higher rates of migration, especially among the younger age cohorts and men.†

10.3 Data Sources

Northeast Brazil has a population of about 56 million spread over 1.5 million km^2. The region is composed of nine states that are divided into 1800 municipalities. Most municipalities have relatively low population levels, with a median of about 12,000 inhabitants. The empirical analysis in this chapter combines information from several sources to build a longitudinal data set of municipalities spanning the 2000–2013 period. Data for a subset of the outcomes of interest are not available for the first or last year of this period, and hence will not be reported or used in the estimation.

Data on agricultural production come from the annual surveys *Produção Agrícola Municipal* (Municipal Agricultural Production, or PAM) of the Brazilian Institute of Geography and Statistics (IBGE). These data are collected every year and contain information on planted and harvested area, physical output, and revenue by municipality and crop (distinguishing between temporary and perennial crops). The PAM covers both rain-fed and irrigated agriculture, but does not distinguish between them. Annual data on livestock and animal-related production by municipality come from *Produção Pecuária Municipal* (Municipal Livestock Production, or PPM), which are available on a yearly basis until 2013. This information was complemented with annual data on municipal agricultural value added and gross domestic product (GDP), also from IBGE. The latter data are fully integrated with the series of standard National Accounts of Brazil and are available until 2012.

* Studies for other countries also point to significant adverse impacts on local labor market outcomes. Focusing on India, Jayachandran (2006) finds that negative shocks to agricultural productivity (induced by negative rainfall shocks) depress wage rates, and shows that these impacts are stronger for workers who are poorer, less able to migrate, and more credit constrained (because such workers have a more inelastic labor supply).

† In a related work for South Africa during apartheid, Dinkelman (2013) shows that short-run migration is more responsive to drought in regions characterized by lower mobility restrictions. Clark and Mueller (2012) use event-history methods and a longitudinal data set from the rural Ethiopian highlands to estimate the impacts of drought on population mobility over a 10-year period. The results indicate that (1) male labor migration unambiguously increases with drought; and (2) land-poor households are the most vulnerable. These results support the hypothesis that mobility serves as a key coping strategy following drought, as well as the common assumption that the poor are most vulnerable to these effects.

To document the evolution of reservoir water volume relative to capacity, this chapter draws on data on the evolution of water storage reservoirs from the National Water Agency (ANA). These data include reservoirs monitored by the agency with volume above 10 hm^3 (cubic hectometers).

Finally, to measure the occurrence of droughts, the study draws on a combination of objective blend indicators from the Northeast Drought Monitor, notably the Standardized Precipitation Index (SPI) and the Standardized Precipitation Evapotranspiration Index (SPEI). These blend indicators are available for each municipality in Northeast Brazil on a monthly basis for the period from January 2000 to October 2014. Two different indicators are considered:

1. *Blend long*: The most intense value considering SPI and SPEI for 12 and 18 months (long-term drought)
2. *Blend short*: The most intense value considering SPI and SPEI for 3 and 4 months (short-term drought)

Each of these indicators was calculated for each monitoring station and interpolated to a raster. Then the average value for each municipality-month was computed.

The first indicator is best suited to capture the impacts of longer term droughts that affect the reservoirs and irrigated agriculture. The second is better suited to capture the effects of shorter term droughts that are likely to affect mostly rain-fed agriculture. The values of these indicators are inversely related with the severity of the drought: more negative values indicate greater water scarcity and hence greater drought severity. A positive relationship between each of these indicators and local economic activity would be expected.

10.4 Empirical Methodology

A first step toward understanding the economic impacts of droughts is to characterize the recent evolution of the main outcomes of interest. This descriptive analysis makes it possible to compare values of key indicators of interest in drought years versus years characterized by normal precipitation levels. Provided that there are no other important factors influencing the outcomes of interest, simple before and after comparisons offer a straightforward and intuitive way of inferring drought impacts. In general, though, it should be recognized that this assumption is a strong one and is rarely valid in practice, especially when economic outcomes such as local income levels are considered. Perhaps the main challenge associated with quantifying the impact of droughts on local economic activity is to determine what would have happened to the indicators of interest in the absence of drought. Although that counterfactual is clearly unobservable in practice, the literature reviewed in this chapter indicates that it is possible to use econometric methods and historical series of municipal-level data to infer it and to attribute certain changes in economic and social indicators to the occurrence of droughts with different degrees of severity. These methods are adopted here.

Specifically, this chapter will estimate the effects of drought on a set of outcomes Y_{mt} in municipality m in year t. The set of outcomes considered includes the logarithms of agricultural production and livestock, animal-related production, agricultural value added, and municipal GDP. As is standard in the literature reviewed in this chapter, it is assumed that outcomes follow a data generating process of the form:

$$\ln Y_{mt} = \beta D_{mt} + \varphi_m + \gamma_t + \mu_{mt} \tag{10.1}$$

where:

D_{mt} is a variable that measures the occurrence of droughts
The parameter β denotes their effect on the outcome of interest

This model includes municipality fixed effects φ_m that capture all time-invariant unobservable differences between municipalities that might affect the economic outcome of interest. For instance, municipalities that are more prone to experiencing droughts may have a different infrastructure that influences the way in which droughts impact the outcomes of interest. Finally, the model also includes year fixed effects γ_t, which control for shocks that are common to all municipalities. These shocks include region-wide changes in economic activity, demographics, aggregate economic policies, and other factors that affect the outcomes of interest. Since the model accounts for municipality and year fixed effects, the coefficients of interest are identified from municipality-specific deviations in drought incidence after accounting for shocks common to all municipalities in the region in a given year. The existence and magnitude of rainfall shortages at this geographical level are assumed to be orthogonal to the error term. In other words, the key identification assumption is that no other shocks to the outcomes of interest are systematically related to municipality-specific droughts. In all econometric specifications, the standard errors are corrected to allow for arbitrary forms of heteroscedasticity and serial correlation by clustering at the municipality level.

10.5 Empirical Results

10.5.1 Trends in Drought Indicators

Tables 10.1 and 10.2 report yearly averages for each different drought indicator considered, for each state in Northeast Brazil and for the region as a

TABLE 10.1

Annual Averages of Long-Term Blend Drought Indicator, 2000–2014

Year	AL	BA	CE	MA	PB	PE	PI	RN	SE	Northeast
2000	−0.41	0.33	0.10	0.42	−0.25	−0.43	0.34	−0.32	0.10	0.06
2001	−0.01	−0.11	−0.23	0.02	−0.01	−0.12	−0.12	−0.11	−0.03	−0.09
2002	0.17	−0.04	−0.15	−0.23	−0.07	−0.11	−0.48	−0.04	0.12	−0.12
2003	−1.08	−1.01	−0.26	−0.48	−0.64	−0.88	−0.74	−0.38	−1.31	−0.74
2004	−0.31	−0.05	0.39	0.12	0.22	−0.08	0.09	0.32	−0.09	0.08
2005	−0.44	−0.29	−0.85	−0.63	−0.51	−0.30	−0.61	−0.49	−0.94	−0.51
2006	−0.44	−0.16	−0.63	−0.32	−0.51	−0.38	−0.45	−0.56	−0.58	−0.40
2007	0.02	0.00	−0.46	−0.40	−0.36	−0.35	−0.43	−0.40	0.13	−0.26
2008	0.12	−0.61	−0.06	−0.10	0.22	−0.03	−0.25	0.14	−0.23	−0.16
2009	0.24	−0.31	0.73	0.72	0.63	0.26	0.50	0.88	−0.37	0.33
2010	0.11	−0.53	−0.56	−0.54	−0.38	−0.12	−0.38	−0.43	−0.45	−0.41
2011	0.37	−0.17	0.10	−0.04	0.29	0.35	0.03	−0.05	−0.18	0.05
2012	−0.82	−1.11	−0.88	−0.82	−0.68	−0.85	−0.98	−0.79	−1.21	−0.91
2013	−1.77	−1.51	−1.69	−1.37	−1.67	−1.71	−1.59	−1.62	−1.92	−1.61
2014	−0.67	−0.75	−0.91	−0.49	−0.72	−0.66	−0.98	−0.68	−0.96	−0.75

Note: Computations based on municipality-month data from the Brazilian Drought Monitor. Data cover the nine states of Northeast Brazil: AL (Alagoas), BA (Bahia), CE (Ceará), MA (Maranhão), PB (Paraíba), PE (Pernambuco), PI (Piauí), RN (Rio Grande do Norte), and SE (Sergipe).

TABLE 10.2

Annual Averages of Short-Term Blend Drought Indicator, 2000–2014

Year	AL	BA	CE	MA	PB	PE	PI	RN	SE	Northeast
2000	−0.31	0.26	0.15	0.32	−0.16	−0.31	0.32	−0.23	0.07	0.07
2001	0.15	−0.03	−0.26	−0.07	−0.05	0.01	−0.17	−0.17	0.11	−0.07
2002	−0.16	−0.22	−0.27	−0.30	−0.31	−0.35	−0.53	−0.24	−0.17	−0.30
2003	−0.88	−0.93	−0.45	−0.55	−0.68	−0.81	−0.81	−0.58	−1.02	−0.75
2004	−0.49	−0.30	0.10	−0.06	−0.09	−0.31	−0.13	−0.01	−0.33	−0.17
2005	−0.53	−0.39	−0.91	−0.68	−0.59	−0.41	−0.69	−0.61	−0.99	−0.60
2006	−0.47	−0.08	−0.67	−0.37	−0.56	−0.44	−0.44	−0.60	−0.52	−0.41
2007	−0.06	−0.10	−0.52	−0.49	−0.34	−0.37	−0.50	−0.41	0.01	−0.32
2008	−0.12	−0.65	−0.34	−0.34	0.00	−0.24	−0.46	−0.07	−0.37	−0.34
2009	−0.04	−0.34	0.48	0.39	0.38	0.06	0.31	0.62	−0.52	0.14
2010	0.06	−0.46	−0.36	−0.35	−0.15	−0.04	−0.19	−0.21	−0.46	−0.27
2011	0.29	−0.15	0.10	−0.08	0.10	0.20	0.02	−0.20	−0.15	−0.01
2012	−0.70	−0.93	−0.83	−0.76	−0.70	−0.78	−0.88	−0.77	−0.92	−0.82
2013	−1.34	−1.24	−1.46	−1.19	−1.32	−1.35	−1.40	−1.29	−1.52	−1.32
2014	−0.58	−0.76	−0.93	−0.50	−0.70	−0.60	−0.98	−0.69	−0.88	−0.74

Notes: Computations based on municipality-month data from the Brazilian Drought Monitor. Data cover the nine states of Northeast Brazil: AL (Alagoas), BA (Bahia), CE (Ceará), MA (Maranhão), PB (Paraíba), PE (Pernambuco), PI (Piauí), RN (Rio Grande do Norte), and SE (Sergipe).

whole. In Table 10.1, the long-term drought indicator suggests that weather conditions leading to drought were particularly adverse in 2012–2013. Between 2000 and 2009, this indicator averaged 0.08 across all municipalities in Northeast Brazil. It then reached a relatively low level in 2010 and the lowest levels in 2012–2013, following a close to average value in 2011. In the first 10 months of 2014, this indicator was again clearly below average, although less so than in 2012–2013. These patterns are fairly similar across states in Northeast Brazil. Table 10.2 reveals that these patterns were also qualitatively similar when considering the short-term blend drought indicator.

10.5.2 Trends in the Volume of Water Reservoirs Relative to Capacity

The relative scarcity of rainfall observed in Northeast Brazil since 2010 (especially since 2012) is reflected in the levels of water available in reservoirs. Following an extended period of below average rainfall, reservoir levels have become dangerously low, placing at risk the ability of communities to maintain adequate drinking water supplies and water for other uses. Table 10.3 reports the evolution of equivalent reservoir volumes in Northeast Brazil from October 2006 to October 2014. The figures presented in this table refer to the reservoirs monitored by ANA with capacity above 10 hm^3 and reveal that the multiyear drought led to sharp reductions in equivalent reservoirs, especially in 2012, 2013, and 2014, with the states of Ceará, Paraíba, and Pernambuco recording especially low levels (i.e., less than 30%).

At least half of the 504 reservoirs monitored by ANA in Northeast Brazil reached the end of 2013 with less than 30% of their capacity. In order to prioritize human consumption, ANA adopted several emergency regulatory

TABLE 10.3

Evolution of Reservoir Volumes, October 2006 to October 2014

Year	BA	CE	PB	PE	PI	RN	Northeast
2006	0.72	0.63	0.85	na	0.72	0.83	0.70
2007	0.74	0.54	0.71	na	0.66	0.72	0.61
2008	0.67	0.78	0.89	na	0.83	0.93	0.80
2009	0.54	0.89	0.91	0.88	0.77	0.94	0.85
2010	0.47	0.64	0.68	0.81	0.73	0.70	0.65
2011	0.42	0.79	0.75	0.76	0.78	0.87	0.75
2012	0.34	0.56	0.49	0.41	0.56	0.59	0.52
2013	0.32	0.38	0.33	0.30	0.42	0.40	0.37
2014	0.45	0.27	0.27	0.24	0.42	0.35	0.31

Note: Data provided by ANA, referring to reservoirs monitored by ANA, above 10 hm^3.

actions that restricted the use of water in some rivers and reservoirs in accordance with Law 9.433/1997 that created the National Policy for Water Resources. These emergency actions ranged from reductions in the exit flow of the reservoir water to establishing specific days for using water from rivers and reservoirs for productive activities, and even to the temporary suspension of water use.

10.5.3 Trends in Main Economic Indicators

This section examines the evolution of key economic indicators before and during the most recent drought: agricultural production, agricultural value added, and municipal income. Table 10.4 reports year-to-year growth rates in the gross value of agricultural production in each state of Northeast Brazil and in the region as a whole, aggregating temporary and perennial crops.

These descriptive results suggest a strong correlation between the drought indicators and the value of agricultural production: in years in which the drought indicators had more negative values (2003, 2005, 2010, 2012, and 2013), the gross real value of agricultural production exhibited negative rates of growth. Furthermore, these losses were especially large in 2012 and 2013, years in which the blend indicators point to especially severe drought.

Table 10.4 reveals that, in 2013, the gross value of agricultural production in Northeast Brazil was about 50% of the level observed in 2009 and 44% of that observed in 2011. It is also evident that the magnitude of cumulative production losses over the 2010–2013 period varied somewhat across states, although it was sizable in almost every one. Tables 10.5 and 10.6 provide similar statistics for temporary and perennial crops, respectively. Both types of crops have experienced marked declines during the multiyear drought, especially in 2013.

Although the data considered earlier measure different components of agricultural production, it is useful to complement this analysis with the evolution of a single and important summary measure of performance in the primary sector: agricultural value added from the IBGE estimates of municipal income. Unfortunately, these data were available only until 2012. Nevertheless, the results are generally consistent with the information reported earlier: agricultural value added declined in the region in 2010–2012, with a particularly large drop in 2012. These losses in agricultural value added were especially large in Ceará (2012), Paraíba (2010 and 2012), Pernambuco (2011–2012), Piauí (2010 and 2012), and Sergipe (2010–2011) (Table 10.7).

The production losses observed in the agricultural sector contributed to the relatively low rates of GDP growth observed in 2011–2012. As Table 10.8 shows, although most states experienced positive rates of real GDP growth over this period, it was generally more sluggish than in previous years and was even negative in Bahia.

TABLE 10.4

Annual Percentage of Change in Gross Real Value of Agricultural Production, Temporary and Perennial Crops, 2001–2013

Year	AL	BA	CE	MA	PB	PE	PI	RN	SE	Northeast
2001	1.73	−4.91	−23.74	15.78	−9.39	8.61	−19.17	−17.51	−6.16	−3.90
2002	−6.87	54.95	50.42	13.04	12.84	16.69	−18.91	81.71	38.34	34.03
2003	−22.78	−16.12	8.18	18.94	12.01	−8.25	109.16	12.70	14.68	−5.74
2004	−3.66	11.47	−15.60	2.70	−10.78	3.70	7.93	3.08	−18.86	3.37
2005	−9.11	−17.74	−13.83	−10.38	−7.34	2.47	−3.94	−14.32	−2.29	−12.32
2006	5.68	1.94	43.67	−10.61	19.02	6.11	−1.15	16.82	29.14	6.81
2007	1.07	23.04	−15.81	6.94	−16.25	−1.15	−13.60	−8.49	−0.42	7.27
2008	15.24	2.53	37.61	68.65	7.33	7.81	89.92	−11.44	20.72	15.28
2009	3.03	−4.61	−25.60	−19.61	2.58	−2.14	−3.19	18.37	18.14	−5.64
2010	−9.44	2.29	−12.25	6.02	−16.04	10.72	−26.76	−11.16	−3.09	−1.44
2011	26.50	11.78	61.91	7.95	26.49	−5.67	77.15	16.00	−18.72	15.27
2012	−9.93	−7.14	−38.35	−4.64	−12.97	−13.23	−5.31	−1.26	2.82	−10.15
2013	−51.24	−53.91	−43.66	−42.31	−46.48	−46.60	−59.68	−51.18	−43.37	−50.66

Notes: Computations based on data from the *Produção Agrícola Municipal*. Data cover Nine states of Northeast Brazil: AL (Alagoas), BA (Bahia), CE (Ceará), MA (Maranhão), PB (Paraíba), PE (Pernambuco), PI (Piauí), RN (Rio Grande do Norte), and SE (Sergipe).

TABLE 10.5

Annual Percentage of Change in Gross Real Value of Agricultural Production, Temporary Crops, 2001–2013

Year	AL	BA	CE	MA	PB	PE	PI	RN	SE	Northeast
2001	1.61	−13.54	−34.69	2.82	−15.77	−2.15	−15.42	−22.29	−11.67	−10.60
2002	−8.47	69.25	65.85	23.26	19.39	30.52	−19.74	105.64	32.25	39.36
2003	−21.82	−15.73	14.18	21.49	13.63	−12.20	127.19	11.14	39.28	−3.41
2004	−3.69	17.26	−23.88	3.49	−8.04	1.76	5.57	0.82	−19.49	4.31
2005	−9.17	−24.85	−10.38	−10.95	−12.20	−0.46	0.04	−18.05	−11.90	−15.84
2006	5.77	−10.29	43.96	−11.34	24.73	1.35	−3.48	23.21	23.62	0.87
2007	1.37	42.21	−13.98	6.45	−19.81	1.02	−11.33	−16.68	15.79	12.30
2008	14.93	6.66	44.78	75.24	8.78	7.16	92.86	−8.27	36.31	22.15
2009	2.93	−15.29	−37.76	−19.85	5.57	6.18	−2.75	23.41	−3.87	−11.21
2010	−9.24	3.60	−15.97	6.44	−16.31	4.85	−26.17	−9.54	22.38	−1.56
2011	27.52	18.87	87.08	8.55	34.02	−1.44	76.80	17.30	−16.50	21.76
2012	−9.90	0.34	−48.14	−4.72	−16.25	−21.73	−3.61	1.14	6.87	−8.46
2013	−51.65	−55.94	−44.82	−41.98	−47.17	−55.29	−60.44	−50.84	−42.84	−52.31

Notes: Computations based on data from the *Produção Agrícola Municipal.* Data cover the nine states of Northeast Brazil: AL (Alagoas), BA (Bahia), CE (Ceará), MA (Maranhão), PB (Paraíba), PE (Pernambuco), PI (Piauí), RN (Rio Grande do Norte), and SE (Sergipe).

TABLE 10.6

Annual Percentage of Change in Gross Real Value of Agricultural Production, Perennial Crops, 2001–2013

Year	AL	BA	CE	MA	PB	PE	PI	RN	SE	Northeast
2001	4.85	9.72	3.84	243.16	12.84	34.43	−35.06	−4.24	−0.97	13.55
2002	33.07	35.84	26.00	−40.69	−4.18	−7.46	−14.35	27.78	43.45	23.12
2003	−39.22	−16.76	−4.32	−8.88	6.76	1.47	15.90	18.35	−4.37	−11.15
2004	−2.95	1.72	4.98	−8.80	−20.25	7.85	31.89	10.81	−18.15	1.00
2005	−7.73	−3.94	−20.04	−0.95	12.02	8.35	−36.22	−2.76	8.38	−3.18
2006	3.86	20.51	43.08	0.20	1.21	14.90	28.46	0.12	34.11	20.22
2007	−5.45	1.35	−19.53	13.46	−2.57	−4.68	−35.34	17.85	−13.89	−2.29
2008	22.53	−4.03	22.05	−12.62	2.73	8.94	51.31	−18.63	3.29	0.28
2009	5.28	14.23	5.73	−13.66	−7.40	−16.28	−10.47	5.46	50.62	9.17
2010	−13.82	0.57	−6.61	−3.66	−15.01	23.37	−37.54	−16.00	−27.06	−1.19
2011	3.42	2.23	27.52	−7.35	−1.81	−13.42	84.80	11.83	−22.22	1.29
2012	−10.72	−18.87	−18.73	−2.18	2.58	4.47	−40.22	−9.39	−4.04	−14.54
2013	−39.81	−49.97	−42.18	−51.87	−42.92	−33.03	−34.40	−52.43	−44.35	−46.07

Notes: Computations based on data from the *Produção Agrícola Municipal.* Data cover the nine states of Northeast Brazil: AL (Alagoas), BA (Bahia), CE (Ceará), MA (Maranhão), PB (Paraíba), PE (Pernambuco), PI (Piauí), RN (Rio Grande do Norte), and SE (Sergipe).

TABLE 10.7

Annual Percentage of Change in Value of Real Agricultural Value Added, 2001–2012

Year	AL	BA	CE	MA	PB	PE	PI	RN	SE	Northeast
2001	5.15	−0.32	−15.62	7.38	1.08	2.54	−0.56	−11.38	5.42	−0.46
2002	−24.58	24.20	11.83	−1.04	−10.40	13.04	−17.63	195.23	8.67	10.10
2003	−18.98	−7.60	7.90	7.24	11.88	2.89	38.10	5.86	33.49	1.61
2004	−5.47	7.14	−12.77	8.41	−14.10	−5.92	−3.43	−6.12	−25.55	−1.05
2005	−6.16	−13.90	−10.37	7.26	−7.11	6.43	−4.30	−14.94	−1.61	−6.13
2006	2.82	−5.12	32.94	3.42	17.92	10.49	−6.34	27.66	21.45	6.30
2007	−9.15	20.52	−11.65	17.27	−17.26	−2.17	−8.95	−14.82	1.18	5.06
2008	13.50	−1.69	22.30	31.01	13.29	14.32	41.55	−10.63	16.73	13.49
2009	2.10	2.44	−21.83	−23.49	2.80	−1.41	4.08	25.60	12.98	−6.43
2010	−2.24	−1.02	−9.25	10.34	−22.13	6.86	−33.34	−12.54	−10.98	−2.60
2011	−2.19	−2.19	18.19	7.97	8.48	−23.16	23.24	−8.72	−22.40	0.12
2012	−10.69	−3.26	−31.57	−9.99	−18.39	−17.30	−44.74	−6.83	17.88	−12.30

Notes: Computations based on data from IBGE's municipal GDP series. Data cover the nine states of Northeast Brazil: AL (Alagoas), BA (Bahia), CE (Ceará), MA (Maranhão), PB (Paraíba), PE (Pernambuco), PI (Piauí), RN (Rio Grande do Norte), and SE (Sergipe).

TABLE 10.8

Annual Percentage of Change in Value of Real Gross Domestic Product, 2001–2012

Year	AL	BA	CE	MA	PB	PE	PI	RN	SE	Northeast
2001	−1.00	−0.48	−1.67	2.11	5.27	1.66	−3.26	2.77	11.11	0.88
2002	1.85	4.61	3.77	1.42	0.98	2.69	1.07	3.90	3.87	3.26
2003	−6.97	−8.53	−8.22	−2.57	−7.27	−9.19	−3.74	−9.77	−6.34	−7.75
2004	5.11	6.08	3.48	6.84	−3.01	2.34	2.23	5.37	2.28	4.04
2005	3.51	8.49	4.79	10.66	5.97	7.04	6.99	8.24	4.14	7.17
2006	9.49	4.36	11.19	11.05	16.27	9.28	12.96	13.08	10.73	9.01
2007	7.52	8.11	3.44	5.09	5.90	6.76	5.19	6.14	6.31	6.39
2008	−1.59	−0.38	7.35	9.47	4.05	1.72	6.59	−0.07	4.04	2.75
2009	7.11	10.83	7.40	1.73	9.79	9.38	11.56	7.58	−0.68	8.18
2010	9.61	6.65	12.25	7.55	5.36	14.96	9.78	9.77	14.67	9.82
2011	7.02	−4.55	4.12	6.26	2.23	1.06	2.79	2.87	0.88	0.83
2012	−2.32	−1.00	−3.34	6.35	3.11	6.06	−1.37	3.35	0.21	1.17

Note: Computations based on data from IBGE's municipal GDP series. Data cover the nine states of Northeast Brazil: AL (Alagoas), BA (Bahia), CE (Ceará), MA (Maranhão), PB (Paraíba), PE (Pernambuco), PI (Piauí), RN (Rio Grande do Norte), and SE (Sergipe).

10.5.4 Econometric Estimates on the Economic Impacts of Drought

This section presents the municipal-level econometric estimates of the average impacts of drought, based on the econometric methodology and data described earlier. Table 10.9 reports the average local impact on the value of agricultural production of yearly changes in each of the three available drought indicators. As noted earlier, the more negative these indicators, the more severe the climate conditions leading to drought. Hence a positive estimated coefficient on the main outcomes of interest would be expected, indicating that a more negative blend indicator is associated with lower agricultural production, whereas a more positive value for the blend indicator is associated with a higher agricultural output.

The coefficients reported in Table 10.9 are indeed positive and tend to be precisely estimated. Since the outcomes of interest are measured in logs, the coefficients exhibit semielasticity between the blend drought indicators and the outcome of interest. For example, the coefficient in column (1) indicates that, if the drought indicator (in this case the long-term drought indicator) changes from 0 to −1 in a given year, the real value of agricultural output is estimated to decline by 20.2%. Similarly, the coefficient in column (2) indicates that, if the short-term drought indicator declines from 0 to −1, the real value of agricultural production is estimated to decline by about 19.4% on average.

These average yearly local estimates can be used to quantify the causal impact of the drought (as measured by the blend drought indicators) on the outcome of interest in the average municipality of Northeast Brazil relative to historical levels of the outcome considered. This long-term drought indicator had an average value of −0.074 over the 2000–2009 period. In the 2010–2014 period, it was always below average except in 2011 (see Table 10.1). Based on the yearly average of the long-term drought indicator

TABLE 10.9

Gross Value of Agricultural Production (log), 2000–2013

	All Crops		Temporary Crops		Perennial Crops	
	(1)	(2)	(3)	(4)	(5)	(6)
Long	0.202***		0.220***		0.027	
	[0.014]		[0.016]		[0.017]	
Short		0.194***		0.206***		0.038*
		[0.016]		[0.019]		[0.020]
Observations	24,316	24,316	24,159	24,159	22,699	22,699
R-squared	0.873	0.872	0.830	0.830	0.921	0.921

Notes: Dependent variable is the log real gross value of agricultural production in the municipality-year over the period 2000–2013. Estimation with municipality fixed effects and year dummies. Standard errors clustered by municipality are in brackets.

***$p < .01$, **$p < .05$, *$p < .1$.

for Northeast Brazil, the estimated losses in the real value of agricultural production would be in the order of 6.7% in 2010, 16.9% in 2012, 31% in 2013, and 13.6% in 2014. In 2011, production would be expected to rise above the normal level by 2.5%. Based on this indicator, the estimated overall impact of the drought over the 2010–2014 period was then about 13% of normal production revenues. For the 2012–2014 period, the estimated loss was 20% of the normal level. It is important to emphasize that these estimates correspond to average losses for the Northeast region as a whole. The underlying data include irrigated and rain-fed agriculture, large and small farms, and different regions. The magnitude of these losses might be expected to be higher for rain-fed agriculture of small producers in the semiarid region.

These estimates are somewhat lower when using the short-term drought indicator, which is more appropriate to capture shorter term droughts. Over the 2000–2009 period, this indicator averaged −0.27. In the 2010–2014 period, it was clearly below the average in every year except 2010 and 2011 (Table 10.2). Using the yearly average values of the blend indicator for Northeast Brazil from this table, the estimated losses in the real value of agricultural production would be in the order of 11% in 2012, 20% in 2013, and 10% in 2014. Production would be expected to stay at about the normal level in 2010 and rise above that level by 5% in 2011. Based on this blend indicator, the estimated overall impact of the drought over the 2010–2014 period was about 7% of normal production revenues. For the 2012–2014 period, the estimated loss was 13.3% of the normal level. Given the duration of the drought in Northeast Brazil, the long-term indicator is likely to be more appropriate to measure its impacts.

The estimates reported in columns (3)–(6) of Table 10.9 indicate that losses in the value of agricultural production caused by the drought mainly reflect losses in temporary crops: the coefficients in columns (4)–(6) are quite similar to those in columns (1) and (2), whereas those in columns (5) and (6) are much smaller. In addition, they are less precisely estimated. However, temporary crops account for the bulk of agricultural production in Northeast Brazil.

The results reported in Table 10.10 concern the relationship between the two drought indicators and headcounts of several types of livestock: cattle, goats, sheep, chicken, and pigs. The econometric results point to statistically significant positive associations between all three indicators and the headcounts of cattle and pigs. For the remaining livestock types, the null hypothesis of no impacts cannot be rejected. With regard to magnitude, the following procedures similar to those outlined earlier, the estimated losses in cattle and pigs associated with the most recent drought, can be computed. Based on the long-term indicator, there were estimated losses of 5% in 2010–2014 and 8% in 2012–2014. Finally, based on the short-term indicator, estimated losses amounted to 3% in the 2010–2014 period and 5% in the 2012–2014 period. For pigs, the estimated coefficients are of similar magnitude, and hence so are the estimated losses.

TABLE 10.10

Livestock Headcount (log), 2000–2013

	Cattle		Goats		Sheep		Chicken		Swine	
	(1)	(2)	(3)	(4)	(5)	(6)	(7)	(8)	(9)	(10)
Long	0.077***		−0.030		0.003		0.037		0.097***	
	[0.009]		[0.034]		[0.032]		[0.023]		[0.024]	
Short		0.087***		−0.056		−0.017		0.028		0.107***
		[0.011]		[0.042]		[0.037]		[0.028]		[0.027]
Observations	24,401	24,401	24,401	24,401	24,401	24,401	24,401	24,401	24,401	24,401
R-squared	0.886	0.886	0.881	0.881	0.847	0.846	0.805	0.805	0.826	0.826

Notes: Dependent variable is the log headcount of livestock of the municipality-year over the period 2000–2013. Estimation with municipality fixed effects and year dummies. Standard errors clustered by municipality are in brackets.

***$p < .01$, **$p < .05$, *$p < .1$.

Table 10.11 presents impacts on animal-related outputs. The results point to a positive association between all three indicators and honey production. For milk and eggs, however, the null hypothesis of no impacts cannot be rejected. Based on the long-term indicator, the drought is estimated to have reduced levels of honey production (relative to the historical mean) by 21% in the 2010–2014 period and 32% in the 2012–2014 period.

Consistent with the results reported in this chapter, Table 10.12 depicts the relationship between the drought indicators and agricultural value added and municipal GDP. Columns (1) and (2) reveal that agricultural value added is significantly reduced when the blend indicators reach higher negative values.

Based on the long-term indicator, the drought is estimated to have reduced value added in the agricultural sector (measured in real terms relative to historical mean) by 3.2% in the 2010–2014 period and by 6% in the 2012–2014 period. Based on the long-term indicator, we obtain estimated losses of 3% in 2010–2014 and 4.7% in 2012–2014. The regressions in which the coefficients are identified include data that go only until 2012. It would have been desirable to include information for 2013 and 2014, but this information is not yet available. With this limitation in mind, the results reported in columns (4)–(6) do not show significant impacts on municipal GDP for the average municipality. This absence of impacts suggests that developments in other sectors (and/or increased income transfers as part of the policy response to the drought) may have contributed to offset the negative effects on local income via agricultural activity.*

TABLE 10.11

Gross Value of Animal-Related Production (log), 2000–2013

	Milk		Honey		Eggs	
	(1)	(2)	(3)	(4)	(5)	(6)
Long	0.006		0.323***		0.049*	
	[0.022]		[0.088]		[0.028]	
Short		0.002		0.404***		0.061*
		[0.026]		[0.106]		[0.032]
Observations	24,401	24,401	24,401	24,401	24,401	24,401
R-squared	0.743	0.743	0.708	0.708	0.751	0.751

Notes: Dependent variable is the log real gross value of animal-related production of the municipality-year over the 2000–2013 period. Estimation with municipality fixed effects and year dummies. Standard errors clustered by municipality are in brackets.

***$p < .01$, **$p < .05$, *$p < .1$.

* Econometric evidence for the state of Ceará suggests that the effects of drought on public finances depend crucially on whether government response is considered (Magalhães and Glantz 1992).

TABLE 10.12

Agricultural Value Added and GDP (log), 2000–2012

	Agricultural Value Added		GDP		GDP—Rural Municipalities	
	(1)	(2)	(3)	(4)	(5)	(6)
Long	0.046***		−0.004		0.088**	
	[0.007]		[0.004]		[0.043]	
Short		0.047***		−0.003		0.094
		[0.009]		[0.005]		[0.058]
Observations	22,647	22,647	22,647	22,647	546	546
R-squared	0.928	0.928	0.989	0.989	0.946	0.946

Notes: In columns (1) and (2), the dependent variable is the log of the real agricultural value added in the municipality-year over the period 2000–2012. In columns (3) and (4), the dependent variable is the log of the real municipal GDP in the municipality-year over the period 2000–2012. In columns (5) and (6), the dependent variable is the log of the municipal GDP in the municipality-year over the period 2000–2012, and the sample is restricted to municipalities where the agricultural value added represented over 50% of the municipal GDP on the average over this period. Estimation is made with municipality fixed effects and year dummies. Standard errors clustered by municipality are in brackets.

***$p < .01$, **$p < .05$, *$p < .1$.

However, when restricting the sample to largely rural municipalities (defined as municipalities where the share of agricultural value in municipal GDP exceeds 50%), there are large and significant losses on overall economic activity. Based on the long-term indicator, there were estimated losses of 5.7% in 2010–2014 and of 8.9% in 2012–2014 (relative to the historical mean).

10.6 Policy Response to the Current Drought: Main Actions and Associated Cost

In recent decades, a number of structural solutions have been put forward to deal with droughts, including water storage and river basin transfer projects. Improvements in water supply infrastructure have also contributed to increase in the region's resilience to droughts. However, these structural solutions are insufficient to overcome multiyear periods of below average rainfall. Like most nations around the globe, Brazil has approached the management of the multiyear drought events that occur every several decades through various emergency relief and response activities. These response actions are often orchestrated by temporary drought committees, led by *Casa Civil* in the president's office at the federal level and by sectoral secretariats at the state level. As indicated in Chapter 1, the emergency actions put forward to mitigate the economic and social impacts of the most recent severe drought have included emergency credit lines, renegotiation of agricultural debts, expansion of social support programs, and tank truck deliveries of emergency drinking water to rural communities. In addition to emergency actions, the policy response to the most recent drought entailed infrastructure developments under the Growth Acceleration Program (PAC), including well drilling, construction of dams, and the provision of equipment to affected municipalities.

It is estimated that R$16.6 billion (close to US$4.5 billion) of federal resources have been allocated for emergency responses and the structural actions associated with the most recent multiyear drought as of 2014.* Of this total, R$7.6 billion (over US$2 billion) corresponds to measures put in place by March 2013. The remaining R$9 billion (more than US$2.4 billion) was from a renewed drought policy package announced in April 2013 by the Brazilian president and minister of national integration, during the 17th meeting of the deliberative council of SUDENE, in which the governors of the states of Ceará, Bahia, Alagoas, Sergipe, Pernambuco, Paraíba, Rio Grande do Norte, Piauí, Maranhão, Minas Gerais, and Espirito Santo also participated. At this time, establishment of the National Drought Emergency Task Force (Força Nacional de Emergência—Seca) and the Drought Observatory (Observatório da Seca) were also announced. The main role of the task force is to monitor the quality of water supply and put forward measures to mitigate the low levels of water in the reservoirs. It is coordinated by the Ministry of National Integration and is also composed of the representatives of Codevasf, DNOCS, CPRM, Hydroelectric Company of São Francisco (Chesf), Bank of Brazil, and ANA. The Drought Observatory is an online portal that gathers detailed information about the emergency and structural policy actions associated with drought.

Table 10.13 shows how that R$9 billion was allocated for various policy actions. One of the most costly actions was debt renegotiation of farmers affected by drought. This measure authorized all agricultural producers in municipalities in the semiarid Northeast of Brazil that were declared to be in state of emergency by the Federal Government to postpone payment of debt contracted during the 2012–2014 period for 10 years. Payments would begin in 2015 for large farms and in 2016 for smallholders. Reduction of debt incurred by 2006, in cases of liquidation of rural credit, was also authorized.

The composition of the R$7.6 billion corresponding to measures put in place by March 2013 was not fully detailed. However, some R$510.1 million (US$134.7 million) was devoted to *Operação Carro-pipa* between January 2012 and March 2013, when the Federal Government hired 4746 water trucks to supply 777 municipalities. Since May 2012, more than 300 operations were also contracted through emergency credit lines for a total of R$2.4 billion (US$645 million).

* It should be noted that the drought has continued in 2015 and 2016, for which federal expenditure data are not yet available. Thus, this total understates the actual amount allocated for this purpose to date.

TABLE 10.13

Resources Allocated to Renewed Drought Policy Package, from April 2013

Action	Federal Resources (R$ Million)
Water truck deliveries of drinking water (*Operação carro-pipa*)	643.5
Re-equipping the army (*Reequipar o Exercito*)	277.9
Production cisterns (*Cisternas de producao*)	640
Drilling and recovering of wells (*Perfuracao e recuperacao de pocos*)	135.8
Producer insurance (*Garantia Safra*)	765
Drought allowance (*Bolsa Estiagem*)	804.1
Subsidized maize (*Venda de milho*)	180
Emergency credit lines (*Linha de credito emergencial*)	350
Debt renegotiation (Renegociacao da divida)	3147
PAC equipment (*PAC equipamentos*)	2100
Total	9043

Note: The sources of these data are the Ministries of Integration and Planning of the Government of Brazil.

There were several important difficulties in quantifying the exact budgetary costs of the policy responses to the most recent multiyear drought. First, the composition of the R$16.6 billion was not fully detailed. Second, this amount includes resources associated with the provision or renegotiation of emergency credit lines to farmers, and it is difficult to quantify what portion will be paid back in the future. Third, some costs refer to measures that are more structural in nature and may therefore have impacts that extend well beyond the current drought. Finally, although the federal programs account for the bulk of the total budgetary cost associated with drought responses, they are generally paired with complementary actions at the state and municipal levels. The latter vary across states and are difficult to map and quantify in a comprehensive way. One of the main tasks of future assessments of the budgetary costs of drought is to gather and examine detailed information on the net budgetary costs associated with each specific measure at the federal, state, and municipal levels.

10.7 Summary and Conclusion

One of the most severe multiyear droughts in decades hit Northeast Brazil starting in 2010. The effects of the drought and the associated emergency response activities are perceived to have come at a high cost to society. However, it has been difficult for decision makers to quantify these costs and impacts. This chapter has contributed to fill this knowledge gap by combining several sources of detailed official data to assess the economic impacts and costs of the most recent multiyear drought in Northeast Brazil.

To measure the occurrence of drought, the study drew on a combination of objective blend indicators drawn from the Northeast Drought Monitor, notably the SPI and the SPEI. These blend indicators are available for each municipality in Northeast Brazil on a monthly basis for the period from January 2000 to October 2014. These indicators reveal that weather conditions were adverse in several states in 2010, and particularly adverse in 2012–2013.

The relative scarcity of rainfall observed in Northeast Brazil since 2010 (and especially in 2012–2013) was reflected in the water available in the reservoirs. Reservoir levels reached dangerously low levels, placing at risk the ability of communities to receive adequate drinking water supplies and water for other uses. At least half of the 504 reservoirs monitored by ANA in Northeast Brazil dried to 30% or less of their capacities by the end of 2013.

Empirical analysis reveals that these conditions had severe adverse impacts on agricultural output. Based on the long-term drought indicator, it is estimated that over the 2010–2014 period, the drought imposed a 13% loss in the gross real value of agricultural output (relative to normal historical levels). For the 2012–2014 period, the estimated loss was 20% of the normal level. These estimates reflect sizable losses in the value of both temporary and perennial crops. The econometric results also point to statistically significant losses of cattle and pigs, as well as some losses in animal-related production, notably honey and eggs.

Like most countries, Brazil has approached the management of the multiyear drought events that occur every several decades through various emergency relief and response activities. The measures taken to mitigate the economic and social impacts of the most recent drought have included emergency lines of credit, renegotiation of agricultural debts, expansion of social support programs, and tank truck deliveries of emergency drinking water to rural communities. In addition to emergency actions, the policy response to the most recent drought also entailed infrastructure investments including well drilling, construction of dams, and provision of equipment. It is estimated that nearly US$4.5 billion of federal resources have been allocated for emergency and structural actions associated with the most recent multiyear drought.

11

Northeast Drought Monitor: The Process

Eduardo Sávio P.R. Martins, Robson Franklin Vieira Silva, Bruno Biazeto, and Carmen Molejón Quintana

CONTENTS

11.1 Information Technology and Data Utilized by the Monitor 143
11.2 The Drought Monitor 144
11.3 Drought Indicators and Support Products 145
11.3.1 Standardized Precipitation Index 146
11.3.2 Standardized Precipitation Evapotranspiration Index 148
11.3.3 Standardized Runoff and Dry Spell Indicators 148
11.3.4 Support Products 150
11.3.5 Layer with Reservoir Conditions 152
11.4 Important Concepts Considered in the Monitor's Design 153
11.5 Operational Arrangements of the Monitor 155
11.5.1 Actors Involved, Responsibilities, and Users 155
11.5.2 Description of the Flowchart of Activities to Update the Drought Monitor 157

11.1 Information Technology and Data Utilized by the Monitor

The Drought Monitor, as has already been discussed in Chapter 4, depends on the collaboration of climate-related institutions and those in the agriculture and water resource sectors from all the northeastern states and the pertinent federal institutions in an effort to identify the degree of severity of droughts in their meteorological, hydrological, and agricultural dimensions. This participatory and collaborative process is working to integrate the relevant state and federal databases in the region, as well as creating a tool based on the combination of different drought indicators and other information products related to this theme. Its objective is to improve understanding of the physical phenomena that determine the duration, progression, or alleviation of a drought situation, refined with information regarding actual drought impacts based on local evidence.

This monitoring tool brings together information from federal and state institutions to produce a single consolidated Monitor map of drought conditions in Northeast Brazil. In the same manner as the institutional cooperation process that underlies it, this map intends to improve the definition and common understanding of the droughts, as well as to increase the efficiency and effectiveness of public policy responses to assist the affected population. The dynamics of its construction seek to integrate the efforts of the various institutions involved, supporting the sharing of information and use of the final product. In this sense, the alternating authorship by different state institutions and coordination by a federal institution reflects a cooperative design, which, by distributing the responsibility for authorship, strengthens the collective vision underlying the project.

There is an evident need for institutional strengthening to ensure access to information, principally in order to guarantee continuous collection of data and expansion of the network. The existing data network is a mixture of conventional and automatic stations, with some of the latter involving automatic transmission and others not. Conventional technologies generally predominate in the state networks, which, together with the automatic ones without automatic transmission, impose a challenge for updating the data generated by the monitoring network. This is because, in order for the Monitor to be produced in a timely way to assure maximum usefulness of its results, data collection should occur in an automatic and rapid manner, which implies the need to overcome some existing limitations such as nonconnection of databases (whether state or federal) with the Internet, and, in some extreme cases, the absence of a database altogether. It should also be noted that there are still insufficient technical staff who are dedicated to drought monitoring (information technology specialists, those responsible for maintenance, communication with observers in the field, data analysis, etc.) and, in more extreme cases, services are outsourced or undertaken by interns for a certain period of time. Investments in the state networks are needed in order to ensure an adequate density of monitoring stations, as well as timely remote access in order to calculate the indicators used by the Drought Monitor.

11.2 The Drought Monitor

The Drought Monitor is not an automatically generated product, but a process that is based on information transparency and the convergence of evidence. Differently from traditional drought monitoring methods, in which institutions and products are treated independently, the Monitor's process, in essence, combines various data sources, products, and information from all of the meteorological, hydrological, and agriculture/livestock monitoring systems of the federal and participating state governments, supported by local information furnished by those who are actually experiencing the drought. Each step of the process is equally important, because the observed data and isolated drought indicators do not necessarily reflect the intensity and/or nature of the drought as experienced in specific parts of the region, for various reasons:

- The low density of the existing data network
- Rainfall totals do not reflect actual impacts on the ground
- Interpretation of different drought information can vary according to the interpreter, in the absence of a single map that takes all the various indicators and support products into account, as no single indicator presents the best results in all circumstances (Heim 2002)

The Drought Monitor must consider the various dimensions of a drought, both in relation to its intensity in terms of the associated impacts and the times cales that affect agriculture, water systems, and the economy in general. On the Monitor Map the regions experiencing different degrees of drought are identified and divided into five categories (D0–D4), represented by distinct colors, that signal the degree of intensity and duration of specific drought events.

The collaborative nature of the initiative between the states and the Union, as well as the local validation process, ensure a consensus among the different levels of government concerning the stage of development of a drought, thereby avoiding disagreements about the criteria used to implement or discontinue actions linked to the National Drought Policy, especially emergency ones. Table 4.1 presented the various categories of drought and their associated impacts as utilized by the Drought Monitor that need to be adapted and contextualized in order to improve the definition and characterization of the droughts in Brazil. It is important to point out that the impacts in question are generic ones that seek to represent, in a simplified way, all the subregions of Northeast Brazil. The Drought Monitor, thus, constitutes a decision support tool with a view toward both preparing for and responding to the effects of the droughts, based on an indication of their severity and likely duration (i.e., short, medium, or long run).

11.3 Drought Indicators and Support Products

It is not possible to create a definition of droughts that functions in all circumstances (Wilhite 2000). One of the reasons is that different sectors, such as agriculture and water resources, depend on different indicators to characterize a drought in a specific area as concerns droughts of shorter or longer duration. This is the main motivation for using different types of indicators in the Drought Monitor process, some calculated from federal and state data networks and others obtained from various national and international institutions. The Monitor differentiates between what is calculated based on observed data, which are called *drought indicators*, and those obtained from some institution, which serve as *support products* for delineation of the map. In addition, the Monitor process is *organic* in that it permits new drought indicators and support products to be incorporated at any time, as long as they add useful information for production of the map. Starting from a good conceptual basis, it is possible to identify inconsistencies between the observed data from the automatic and conventional monitoring stations and the local impacts in various areas in designing the map, as well as to understand and describe the climate variations that may have been responsible for these impacts in the areas considered.

At present, the Drought Monitor is based on three drought indicators, two of which are meteorological that focus on the short and longer run, respectively, and the third, hydrological, which is only short term. All the drought indicators of the Monitor are calculated based on historical records from the monitoring stations, which also is the case with most of the support products. The following paragraphs provide a brief description of the information utilized by the Monitor.

11.3.1 Standardized Precipitation Index

The standard precipitation index (SPI) was developed by McKee et al. (1993) and is considered an indicator that is easy to obtain because it only uses rainfall as its entry data. From observed data from all the conventional and automatic rainfall monitoring stations, it is possible to determine the SPI for each station for 3-, 4-, and 6-month intervals in order to identify possible short-term droughts, as well as for intervals of a year or more (12, 18, and 24 months) to identify longer term ones. These intervals could seem arbitrary, but these typical periods of rainfall deficit are related to the types of agricultural and hydrological impacts in a region. Each one of these historic series of precipitation data (3, 4, 6, 12, 18, and 24 months) is adjusted by a probability distribution* (gamma distribution) to identify the probability of the occurrence of a given amount of rainfall during each of these periods. Once this is established on the basis of historical records, the probability of any precipitation to be observed can be calculated through the use of the inverse of a normal distribution pattern (mean = 0 and standard deviation = 1) and it is possible to calculate relative deviations from the mean values for these 3-, 4-, 6-, 12-, 18-, and 24-month periods. The resulting figure is the SPI for a specific value of precipitation during these periods. Negative values indicate rainfall below the median (dry periods) and positive value levels above the median (wet periods), since drought diverges from the normal state of regional precipitation.

---

* A probability function is a function that assigns probabilities to the values of a random variable, here the n-months precipitation ($n = 3, 4, 6, 12, 18, 24$).

The ideal situation would be to work with a 30-year or more time series of continuous precipitation data in order to better characterize the rainfall distribution in the period considered, thereby permitting a better evaluation of the recurrence of a current indicator value in relation to the larger historical period. However, due to data availability limitations, the time series utilized is for an average of 25–30 years for the conventional federal stations and 10 years for the automatic state ones.

As what is of interest is to represent dry periods, only the negative values of the SPI are considered in the Monitor. The drought classification is made in percentiles, such that lower values are related to events with a larger number of occurrences and higher ones with those of lesser occurrence, as can be verified in Table 11.1. An example of an SPI map for the short run can be observed in Figure 11.1.

Among the advantages of the SPI are the following: (1) flexibility of its use over different time scales that reflect the drought's impact on different water sources; (2) the indicator possesses a single value of a probabilistic nature (historical context of the rainfall meter data); it is a *relative drought* concept, which will be discussed next; and (3) it is spatially consistent, permitting comparisons among different localities. In terms of its disadvantages, (1) precipitation is the only parameter used, so that the drought is evaluated only in terms of this variable; (2) it does not include other components of the hydrological balance; and (3) the values change when the data series is updated.

TABLE 11.1

Drought Stages or Categories Which Define the Intensity of the Drought on the Monitor Map Associated with the SPI/SPEI Indicator Values

Category	Percentile	Description	SPI/SPEI
D0	30th	Abnormally dry	−0.5 to −0.7
D1	20th	Moderate drought	−0.8 to −1.2
D2	10th	Severe drought	−1.3 to −1.5
D3	5th	Extreme drought	−1.6 to −1.9
D4	2nd	Exceptional drought	>−2

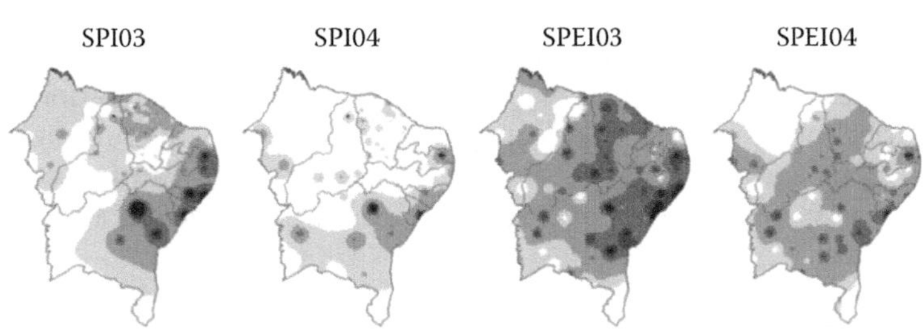

FIGURE 11.1
Examples of drought intensity levels according to the SPI and SPEI for three and four months.

11.3.2 Standardized Precipitation Evapotranspiration Index

The Standardized Precipitation Evapotranspiration Index (SPEI) was developed by Vicente-Serrano et al. (2010) and is calculated with data from meteorological stations that measure both precipitation and temperature. It is similar to SPI in the way in which it is estimated, but calculates a simplified water balance, weighing rainfall and evapotranspiration information, including the effects of temperature variability, for purposes of drought evaluation. The evapotranspiration rate can be calculated on the basis of different formulations. This indicator is also calculated for short-run intervals of 3 and 4 months and of 12, 18, and 24 months for the long term, with 6 months considered as an intermediate value. An example of this indicator can be seen in Figure 11.1.

Highlights of the use of SPEI are (1) flexibility of use for different time scales, just as with SPI; (2) the indicator also has a single value and is of a probabilistic nature (historical context of precipitation and temperature data); (3) it is spatially consistent; and (4) possesses a component of water balance in the soil. Its principal disadvantages are (1) the values change with updating of the data series; (2) it is very difficult to obtain long-term data series for both temperature and precipitation for the same locality; and (3) it is sensitive to the calculation of potential evapotranspiration.

11.3.3 Standardized Runoff and Dry Spell Indicators

The need to include hydrological indicators in the Monitor resulted in the analysis of several options, from the direct use of data from hydrological monitoring networks (reservoir levels, streamflows, etc.) to the use of precipitation characteristics related to rainfall concentration and the occurrence of *dry spells*, which refer to short periods of summer-like climate characteristics at other times of the year. Given the known problems with streamflow data series from the national stream gauge network, it was decided to utilize indirect indicators that could be calculated for each rainfall monitoring station in order to facilitate the design of the Monitor Map. There is also a need to represent the agricultural drought, and an indicator based on dry spells would well represent such a drought.

The variability of both discharges and rain-fed agricultural yield are largely explained by certain key characteristics of the precipitation process, rainfall concentration and the occurrence of dry spells, as mentioned earlier. The drainage discharge is closely connected with rainfall concentration, and for this reason, the corresponding indicator seeks to reflect this concentration:

$$I_e = \sum_{i=1}^{n} L_i W_i$$

where:
L_i is the duration of the ith wet period
W_i is a function of L_i ($W_i = 1$ if $L_i < 10$, and $W_i = 5$ if $L_i >$ or $= 10$)

Rain-fed agricultural output can also be affected by rainfall concentration due to the possible flooding of some areas, so this indicator can explain part of its variability. The length of the wet period was defined here as three or more consecutive days with precipitation exceeding 10 mm. A greater weight is attributed to wet spells that last at least 10 days, and these weights are calibrated with specific regional data.

An indicator related to the occurrence of dry spells during a given period, in turn, is a reflection of both the generation of drainage outflows and rain-fed agricultural output. The indicator of *dry spells* utilized here was defined as

$$I_v = \sum_{i=1}^{n} L_i W_i$$

where:
L_i is the duration of the ith dry period
W_i is a function of L_i ($W_i = 1$ if $L_i < 10$, and $W_i = 5$ if $L_i >$ or $= 10$)

In this case, L_i is defined as three or more consecutive days with rainfall less than 2 mm. A greater weight is attributed to dry periods that last at least 10 days due to the high losses associated with long periods without precipitation.

Calculation of these indicators is undertaken for three- and four-month periods, with adjustment of the historical series of I_e and I_v by a gamma distribution in order to define the probability of not exceeding a specific value of these indicators. Once this is established on the basis of historical records, the probability of not exceeding a particular value of the observed indicator can be calculated and, from the use of the inverse of a normal distribution pattern (mean = standard deviation = 1), it is possible to calculate relative deviations of indicators in relation to the mean for three- and four-month periods. These indicators were denominated standardized runoff indicator (SRI) and standardized dry spells indicator (SDSI) and are visualized in the scale used in the Monitor. An example of the SRI for three months can be visualized in Figure 11.2. At present, due to the need for daily data for the calculation and its availability by the states at an operational level, the SRI and SDSI are being calculated only for a three-month period for Bahia, Ceará, Pernambuco, and Rio Grande do Norte.

Among the advantages of the indicators described earlier are (1) simplicity; (2) consistent with the runoff process; (3) avoid the use of direct flow and reservoir levels data due to the problems already recognized, that is, influence of the reservoirs, inconsistencies in the data series, and absence of or problems with the liberation of water from the reservoirs in the majority of the states; and (4) use of temporal daily rainfall distribution characteristics that are related to runoff.

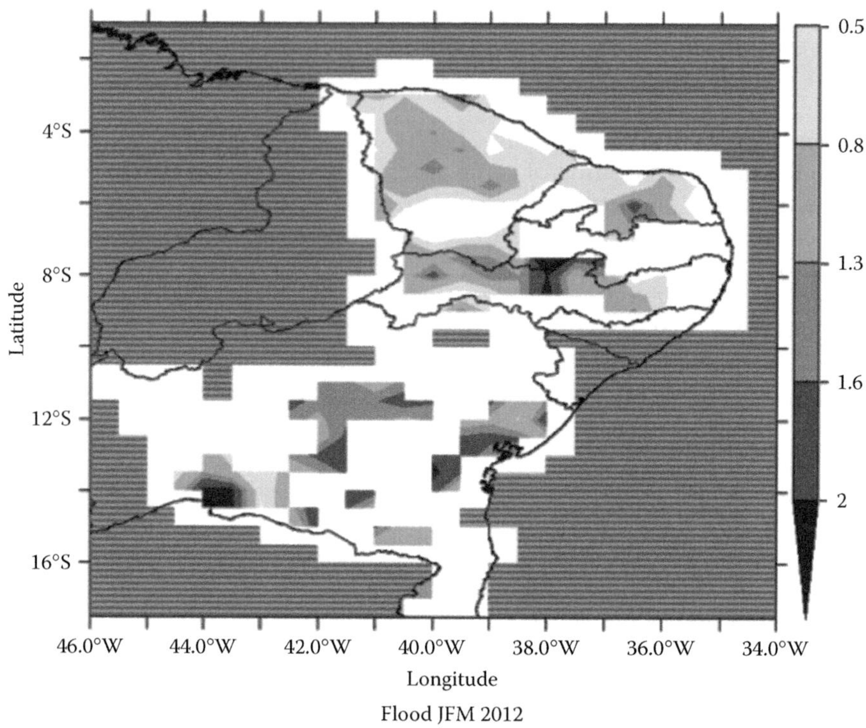

FIGURE 11.2
An example of the standardized runoff indicator for the January–March 2012 trimester. The colors represent drought intensity according to the Drought Monitor's categories. The areas in white represent the without-drought situation and those in hatched-grey lines, the lack of information.

The disadvantages of these indicators, in turn, are due to the difficulties of monitoring the past and the present, such as the needs for (1) a long time series of daily rainfall observations to adjust and calculate the parameters for each monitoring station; (2) daily observed outflow data for validation purposes; and (3) need for a dense network of observational stations for better representativeness for data interpolation.

11.3.4 Support Products

As observed earlier, the support products are information obtained from national and international institutions that help the delineation of the Monitor Map and complement the information provided by drought indicators calculated from data derived from the federal and state monitoring stations in the Northeast region. They are especially important in areas where there is a low density of such stations. The support products used to the present time are (Figure 11.3)

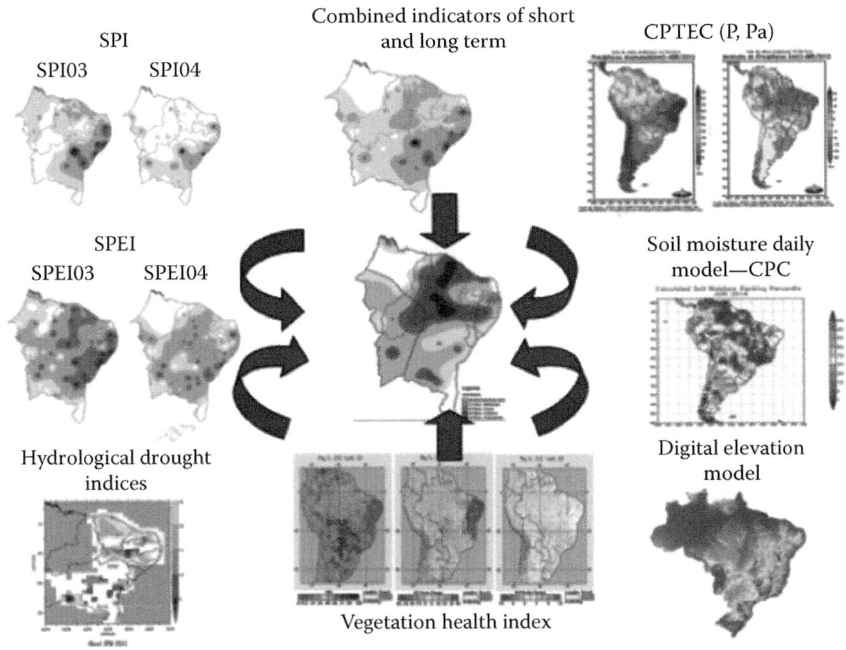

FIGURE 11.3
Representation of the products used in the design of the Drought Monitor Map, indicators, and support products based on remote sensing.

- Layers with composite short- and long-term indicators, organized so as to represent droughts at different time scales
- Accumulated short- and long-run precipitation, climatological norms, SPI (calculated only with federal data), rainfall deviations, and precipitation classified by quantiles from the National Meteorological Service (INMET)
- Accumulated short- and long-term precipitation, climatology, and anomalies from the Weather and Climate Prediction Center (CPTEC)
- Precipitation combined from observed data and estimates by the Tropical Rainfall Measuring Mission (TRMM) of NASA based on a routine by CPTEC
- Soil humidity by CPC/NOAA
- Index of vegetation health (VHI) from the Satellite Applications and Research (STAR) of the NOAA Satellite and Information Service (NESDIS)
- Digital elevation model (MDE)
- Layers corresponding to the previous month's Monitor Map

11.3.5 Layer with Reservoir Conditions

The situation of the reservoirs is essential information in the determination of drought impacts, and is closely related to the operational drought, mainly when the existing information refers only to the reservoir level or volume without a relative view of what occurred over the historical period. The majority of the states do not have these historical data, which should include information on water releases and reservoir operation. This makes it difficult to use such absolute value information for the Monitor since a drought condition is relative to what normally occurred over the history of a particular area. On the other hand, the need to provide information regarding the reservoirs is unquestionable and, for this reason, it is incorporated in the Monitor in a complementary way that represents the physical drought in an auxiliary layer.

The approach suggested is to utilize geometric figures to represent the strategic reservoirs in the Northeast region, whereby the criticality of the reservoir in terms of the degree of restriction on the use of its water for human consumption and irrigation is indicated in the colors associated with the Monitor's categories. In Figure 11.4, it is possible to verify how this layer was operationalized for the state of Pernambuco. Each reservoir is represented by

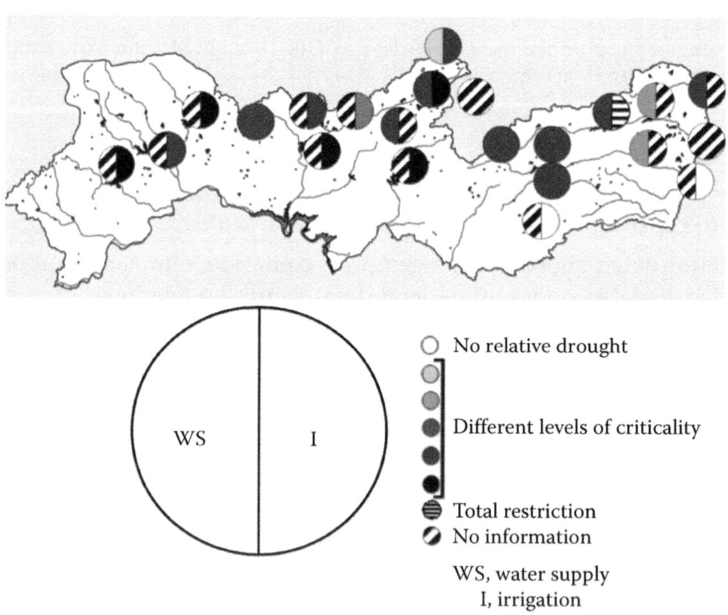

FIGURE 11.4
Classification of drought severity in relation to reservoir water uses (water supply to the left and water for irrigation to the right). White represents a nondrought situation, the gray scale represents the levels of criticality associated with the colors of the Monitor, and horizontal stripes indicate total restriction, while diagonal stripes refer to lack of information.

a circle divided in the middle to represent the two conditions of use considered (on the left, human consumption, and on the right, irrigation).

11.4 Important Concepts Considered in the Monitor's Design

For a better understanding of the process of creating a map from the Drought Monitor, relevant theoretical aspects should be highlighted, as well as the solutions that were obtained during its experimental phase, which were later consolidated based on the principle to start simple and learn by doing. It should also be emphasized that, even though the Drought Monitor for the Northeast region was based on that for the United States, several adjustments were necessary so that the U.S. version could be implemented in Brazil, mainly by incorporating the experience of federal and state professionals engaged in drought-related questions and respecting the present information network in the country.

The experimental phase was initiated following training that took place in August 2014 under the heading of "Methodology for the Design of Drought Monitor Maps for the Northeast" with the objective of producing experimental monthly maps by simulating the Monitor's operational production process. During the experimental period, the operational platform of the Monitor, which is described later in this chapter, was built. Map production activities occurred in a gradual way during the first month and were optimized as the authors and validators gained experience working together. As each month passed, additional information was incorporated, the number of participants grew, and greater agility was achieved in updating the map. Among the lessons learned, it is important to highlight the involvement at both the federal and state government levels. The authors displayed an extraordinary effort through their joint work as well as strong commitment. Meteorologist Adilson Gandu's statement translates this sentiment: "Other maps with SPI type indices exist for Brazil. But the difference is that our map was put together by many hands, eyes, and brains. It is not just the result of an equation."

The first concept that merits highlighting is that of *relative drought*. The indicators used by the Monitor are calculated considering the historical data provided by each monitoring station and obtaining a single value of a probabilistic nature. This permits comparisons to be made among different localities, making it possible to analyze the drought situation throughout the entire northeast region in a spatially consistent manner. As a result, the situation in a particular locality is being assessed in relation to its historical record and not necessarily in relation to the drought perception, which is subjective. As an example, the Northeast *sertão* for a particular month can be classified in the Monitor as being in a less severe drought stage compared with a coastal region (in relation to the categories presented in Table 4.1); even though the *sertão* is, in its essence, drier than the coastal zone, the representation indicates that the latter is in a more advanced stage of the drought in relation to the observed impacts than the former.

Another very important consideration is the type of information the Monitor should provide, that is, the drought as given by the natural conditions inherent in the physical phenomena or that reflecting drought management processes. As with the examples adopted in the other countries where the Drought Monitor was implemented, the institutions that participate in the Brazilian Monitor have agreed that the focus should be on the *natural or physical drought*. The principal justification is not to consider systems that are subject to human management so as to avoid possible conflicts of interest and questioning that could undermine the credibility of the Monitor. Drought management will be addressed by contingency or drought preparedness plans designed by specific systems, as described in Chapters 5 and 12. An example of information that is not considered in the Monitor's design, therefore, is the condition of the reservoirs, which is presented in an auxiliary layer, as described in Figure 11.4.

Since the first Monitor Maps were produced in an experimental form, it was recommended that the region should be considered as a whole on the map, avoiding the use of political–administrative divisions. It is natural that the authors and validators feel more secure in analyzing their own states, and there is also more information for some subregions than others. In addition, neighboring localities commonly present different opinions about drought conditions in their adjoining border areas. This is a particularly delicate question because, in some cases, both neighbors base their views on solid arguments, but there is a need to reach a consensus. Again, it is important that this consensus avoids giving preference to political–administrative divisions, which could be seen as a case of favoritism, considering that the Monitor may be linked to one or more contingency or drought preparedness plans.

During preparation of the map, impacts in different sectors such as agriculture and water resources, which are associated with short- and long-term drought influences, respectively, were considered. The most intense drought category should normally be prioritized and indicated with the so-called *impact lines* identified by the Monitor by L (long-term drought), S (short-term drought), or SL (short- and long-term drought).

The density of the monitoring network, together with the need to combine different indicators, means that the Monitor's design will not be perfect, there being an evident limitation on the spatial representation of the drought that needs to be understood and considered. The Monitor provides a macro view of the situation, but, as it considers information from various sources and people, it can be considered as the best consensus with respect to the stage of the drought's severity that can be obtained. At the present stage, it is still very difficult for the Monitor to represent the drought at the municipal level, but it does present current meteorological, hydrological, and agricultural/livestock conditions, keeping in mind that these are always compared with respect to the historical situation in each area.

11.5 Operational Arrangements of the Monitor

11.5.1 Actors Involved, Responsibilities, and Users

The *actors* responsible for the monthly updating of the Monitor Map are (1) institutions that provide the data; (2) the Central Institution (CI); (3) authors; and (4) validators. In addition to those who are responsible for updating the Monitor Map, the persons and institutions that make use of the results of this process should also be highlighted, together with a group that was created to evaluate the data and indicators utilized by the Monitor, called the *Technical Group*. Each actor possesses responsibilities within a sequence of activities that are summarized next in the order in which it intervenes in the process:

- The institutions that provide data are all the federal and state entities, together with some international ones, that furnish information to update the Drought Monitor. This information can take the form of observed data, which are used to calculate the drought indicators, or products that can be utilized to support the Monitor's Map design in some way.
- The CI is the institution responsible for directing the updating activities and development of the Monitor. This function will be exercised by the National Water Agency (ANA) due to its position as a national agency and the need for the Monitor's compatibility with federal policies, which must be continuous and consistent with the concept of a regulatory agency, as exercised by ANA. The operational functions of the CI are as follows:
  1. Compile the observed data from all federal and state institutions possessing information of interest to the Monitor
  2. Format and control data quality
  3. Calculate the drought indicators used by the Monitor
  4. Obtain the support products from national and international institutions
  5. Prepare the project in a Geographic Information System (GIS) that will be utilized by the author to produce the map. GISs are important tools to collect, store, recover, transform, and represent spatial data in the real world (Burrough and McDonnell 1986)

6. Interact with the authors of the Monitor in the map elaboration process
7. Facilitate the Monitor's validation process
8. Publish the final Drought Monitor Map and maintain the internet site
9. Issue the Monitor's Annual Report
10. Train the authors and validators.

- The *authors* are the institutions responsible for representing the present situation of the natural/physical drought on the Monitor and coordinating the validation process. They are composed of professionals who are involved in the technical work, avoiding political bias and working in an impartial manner. For this purpose, they utilize scientific information (drought indicators and support products) and a GIS to create polygons indicating areas that present different degrees of drought severity. In the initial stage of the Monitor's operation, the authors were institutions in the states of Bahia (Environmental and Water Resources Institute of Bahia, INEMA-BA), Ceará (Ceará Foundation for Meteorology and Water Resources, FUNCEME-CE), and Pernambuco (Water and Climate Agency of Pernambuco, APAC-PE), that does not impede new authors being incorporated in the future, principally if the Monitor Map is updated more often than once a month.
- The *validators* are the institutions responsible for informing the local situation to confirm the accuracy of the proposed map based on the drought indicators presented by the authors. They are directly experiencing the drought and can verify conditions on the ground. The validators are institutions in different sectors connected with the drought from all the northeastern states. The validation exercise should preferably be accompanied by solid evidence-based arguments to avoid subjectivity.
- Once produced and validated, the results of this process need to be utilized and reach all levels of society. The Drought Monitor, accordingly, will be in the public domain and can be accessed by users in general as well as institutions that make use of the results of the process in order to make decisions. It is thus important that the Monitor employ simple language but also meet the needs of the technical user with outputs directed toward the different sectors affected by the drought. Appropriation of this tool by the society as a whole is the greatest guarantee that its transparency and longevity will be maintained.
- *Technical Group for data quality and IT.* In August 2014, a technical group composed of professionals was established to support calibration of the drought indicators, data consistency, and verification of remote sensing data use. The state of Rio Grande do Norte, represented by

the Rio Grande do Norte Agriculture Research Center (EMPARN), leads the Group, which is composed of the authors of the Monitor, ANA, INMET, National Center for Risk Management and Disaster (CENAD), and universities.

11.5.2 Description of the Flowchart of Activities to Update the Drought Monitor

In order for the Drought Monitor's Map to be updated, a specific monthly routine is followed. This sequence of activities is illustrated in Figure 11.5, which presents the Monitor's operational scheme in a simplified manner. These activities can be divided into four steps: (1) data preparation; (2) Monitor design; (3) Monitor validation; and (4) map publication.

Step 1, data preparation, is carried out by the CI and is crucial for the entire process of map design. The CI compiles the data from all the pertinent federal and state institutions in Brazil, in order to calculate the drought indicators, SPI, SPEI, and SEI. It is important to utilize as much information as possible, which is the reason for incorporation of the support products (obtained from national and international institutions) that function as auxiliary data and contribute to the

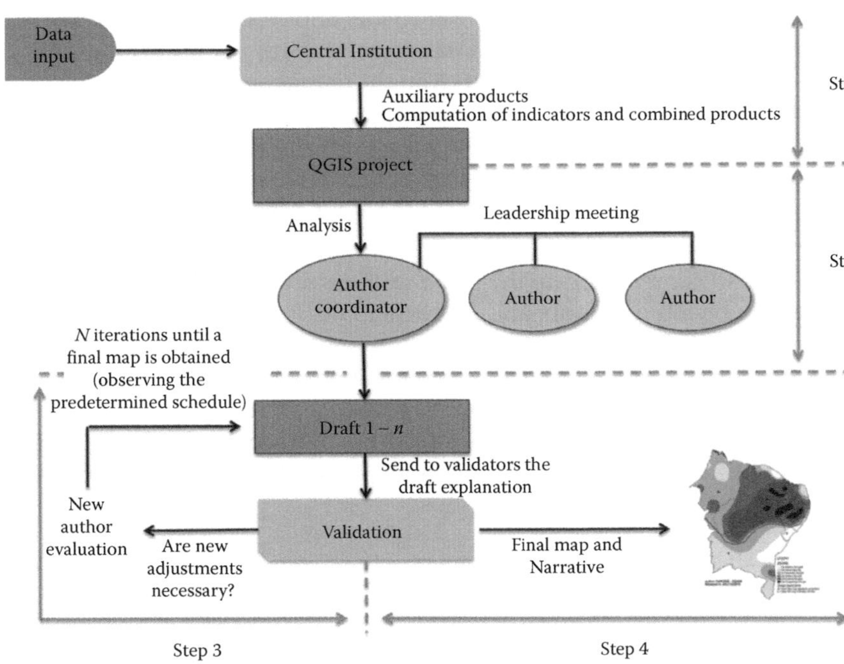

FIGURE 11.5
Operational scheme for the Drought Monitor.

convergence of the evidence, especially where the data network is sparse and/or when there are different outcomes among the various drought indicators. The drought indicators, associated maps, and support products are prepared by the eighth day of each month, formatted, and made available to the authors in a GIS program. All the information is sent to a list of emails comprised of the authors and CI, including the chronogram with the due dates for the design and validation of the Monitor. This chronogram is prepared at the beginning of each year for all 12 months and indicates which author, and consequently which state, is responsible each month for data preparation, design, validation, and publication, leaving aside weekends and holidays.

Step 2, or the process of authoring the map, is the responsibility of the author charged with representing the current drought situation (or absence thereof) in the region. The starting point is the previous month's map. Droughts set in and evolve in a slow and gradual manner, and this needs to be represented. There should also be a minimum of discontinuity in the Monitor's categories as the months pass. The author analyzes the updated drought indicators and support products provided by the CI, and prepares the first version of the map (Draft 0—R0*) together with a text explaining the key elements that are reflected in R0's design. In this stage, precipitation levels during the preceding months and soil and vegetation conditions are verified, as well as the areas experiencing short or long-term drought, keeping in mind that drought management information is not entered in the map. The accompanying text is refined during the entire process and is presented together with the Final Map, at which point it is referred to as *the Narrative*. This process is normally completed within two days after the data prepared by the CI are received.

The R0 is presented to the other authors and discussed in a meeting in which the authors and CI participate, known as the *authorship meeting*, which occurs once the draft map is finalized. This meeting occurs by video conference and has helped the authors to engage in a continuous learning process and to maintain the techniques developed over time. In addition, it has helped to overcome possible climatological, geographic and physical knowledge gaps, as

* R0: initial draft of the Monitor made by the author based on the current drought indicators and support products in addition to the previous month's map, which includes the drought categories and associated impacts. Professional experience, geographic and climatological knowledge, current climate situation, and evolution of the drought during the previous months are also reflected in its elaboration. It represents only the physical/natural drought, not the management systems, such as the reservoirs. It does not have local validation even though the author can include impact information of which it is already aware.

other authors can contribute to the map's design by highlighting conditions and situations in a particular part of the region that are unknown to the current author. As the authors rotate over time, this meeting is also important to keep everyone well informed as to the new Monitor Map that is being prepared. Following the authorship meeting, the author makes the necessary adjustments and prepares a new draft, R1 (see an example of R1 in Figure 11.6), which is then sent to the validators together with other documents that are jointly referred to as validation INPUTs.

Step 3 corresponds to validation of the map through an iterative process between the author and the validators, coordinated by the author, and passing through various drafts before reaching the final

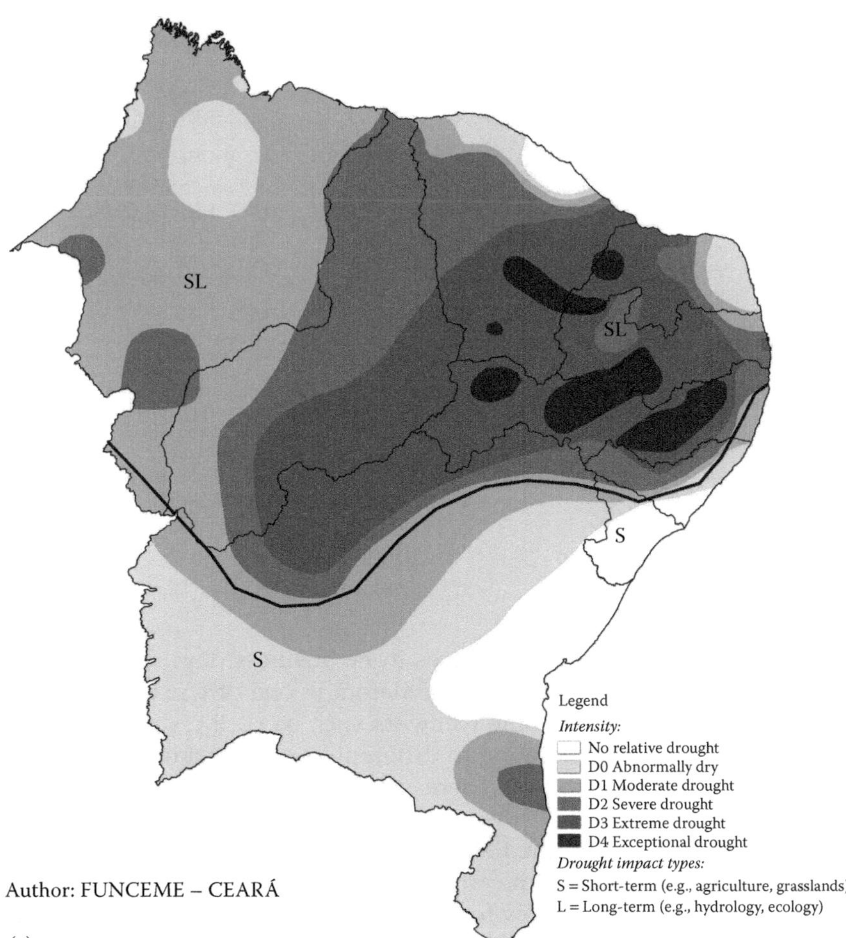

FIGURE 11.6
Drafts R1 (a) of the Monitor's map. *(Continued)*

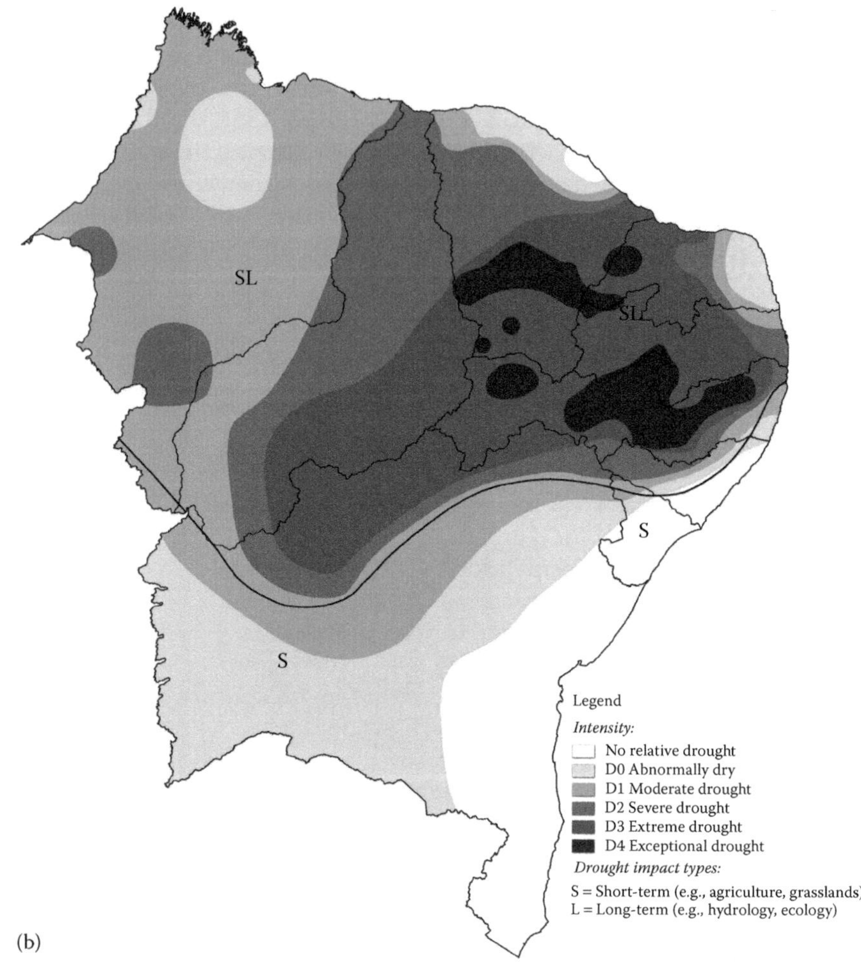

FIGURE 11.6 (Continued)
Drafts R2 (b) of the Monitor's map.

version, which normally requires three business days. The relation between the authors and the validators is one of trust, developed over time, respecting some elements such as (1) the validators can present their argumentation in different levels of detail and knowledge; (2) in some situations, there will only be a report without proof in the form of data; (3) the author needs to confirm the information that comes from the field because it is responsible for the map that will be published; the engagement in the process will give the authors experience in how to interact with each validator; and (4) as it has ultimate responsibility, the last word will be that of the author, even though the dialogue between the author and validator should always be encouraged.

The validation process starts when the author sends the following validation INPUTs (author → validator): (1) the R1 figure; (2) the text explaining the design of R1; (3) the validation form;* (4) a link for primary products that were used to construct R1 (indicator maps, support products, etc.); (5) the validation chronogram; and (6) a spreadsheet with the condition of the priority reservoirs in each state, as well as their names, locations, and water levels. The validator analyzes R1 and agrees or disagrees with the suggested drought categories, based on local information. Whether it agrees or disagrees with the draft, the validator responds with its suggestions for alterations and/or comments on the validation form. This response is considered the validation OUTPUTs (validator → author), which includes (1) the filled-in validation form, indicating whether the validator agrees or not with the map design, explaining the reasons for this agreement or disagreement based on solid argumentation to avoid subjectivity; and (2) the filled-in spreadsheet with the condition of the reservoirs in terms of their use restrictions for irrigation and water supply for human consumption (water resource validators).

After receiving the validator's comments on R1, the author analyzes them and prepares a new draft, R2 (Figure 11.6b). The revised map is sent again for validation and the procedure to send INPUTs and receive OUTPUTs is repeated until a consensus is reached that permits production of the Final Map (Figure 11.7) and Narrative.

After finalizing the iterative validation process, the last phase of the production of the Monitor, *Step 4*, is initiated. The author sends the material to be published to the CI: the Final Map together with the Narrative. The Narrative is divided into two parts: (1) description of the climate conditions; and (2) synthesis of the map design. The second part is described in a general way (i.e., for the entire northeast region) and divided by state to facilitate visualization by the final users. The Narrative is an important tool that should be utilized to the maximum extent possible with the author providing all information considered to be of relevance in the text. The main interactions between the author and validators, climate information, associated impacts, evolution/diminution of drought-affected areas and issues related to production of the map should be incorporated in the Narrative, as well as management information that is not contained in the Monitor, such as the reservoir conditions, for example.

* The is the form to verify which validators evaluated the map to standardize the format for responses and facilitate information storage. In the form, the validator has available the draft, on which it can suggest modifications directly on the map, as well as answer several multiple choice questions and make other comments. The arguments, information, and data utilized to support suggested modifications should be annexed to the form.

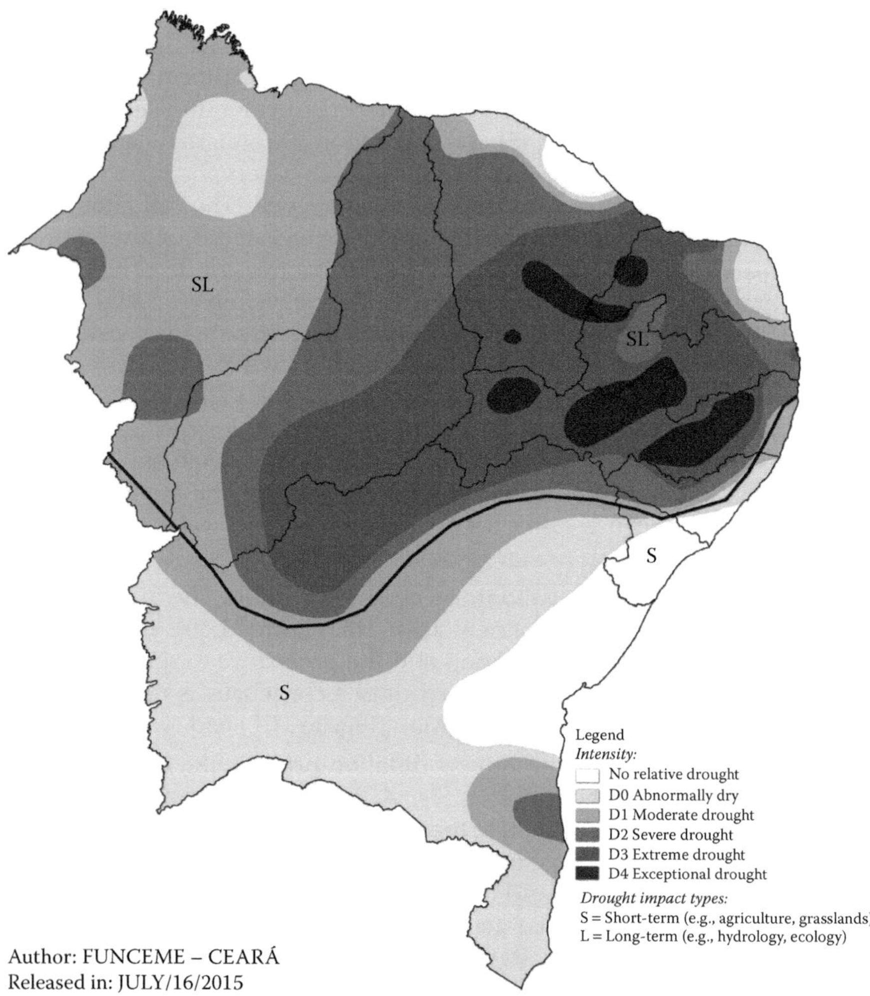

FIGURE 11.7
Final Map of the Monitor published in July in reference to the June 2015 monitoring.

Publication of the Monitor's results and the auxiliary visualization products* is the responsibility of the CI, which will house a link on its website with the following content: (1) the systematized Final Map and its Narrative (example in Figure 11.8); (2) the drought indicators utilized; (3) the support products utilized; (4) information

* See Step 1 of the Description of the Flowchart of Activities to Update the Drought Monitor. These auxiliary visualization products are support products obtained from national and international institutions that are provided by CI in order to help the drawing of the first draft of the map (R0).

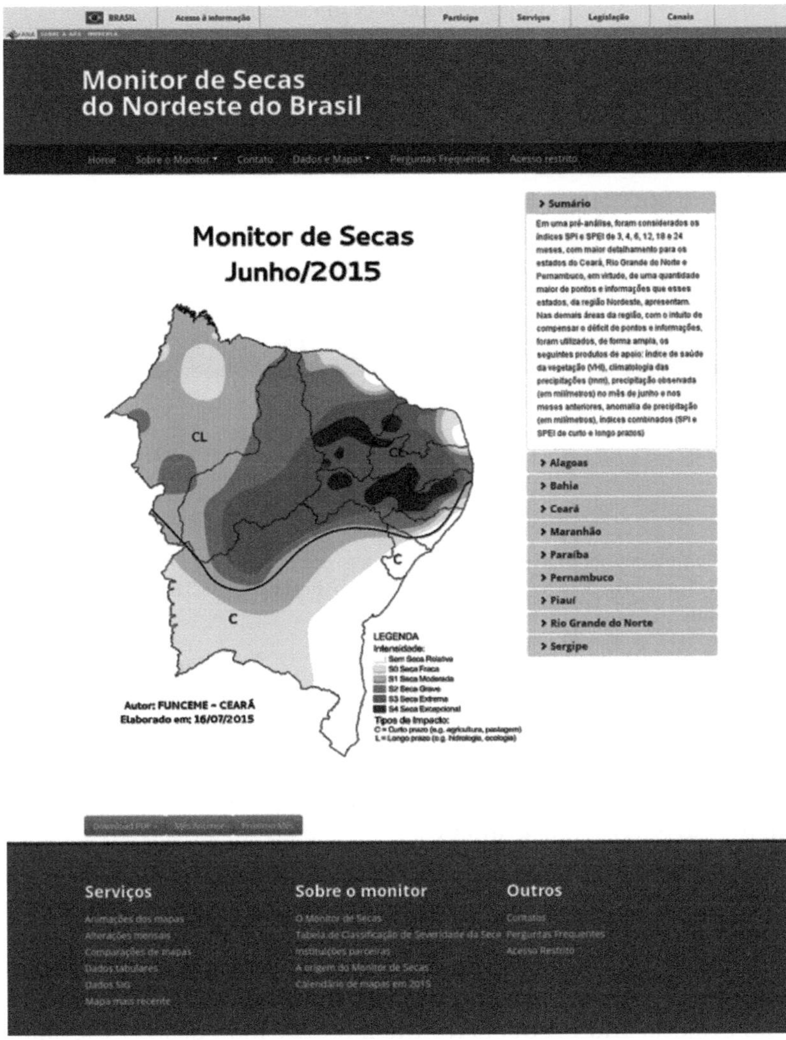

FIGURE 11.8
Layout of the Monitor site: Monitor Map with the physical drought and Narrative. The Narrative is separated in a general part and another detailed by state. (From http://monitordesecas.ana.gov.br.)

about condition of the reservoirs; (5) map animations and comparisons; (6) variations of categories among maps of the Monitor by region and states (Figure 11.9); and (7) maps of changes in drought categories in 1-, 2-, 3-, 6-, and 12-month intervals (example in Figure 11.10) and for reference periods corresponding to the beginning of the calendar year (January 1st) and of the hydrological year (October 1st).

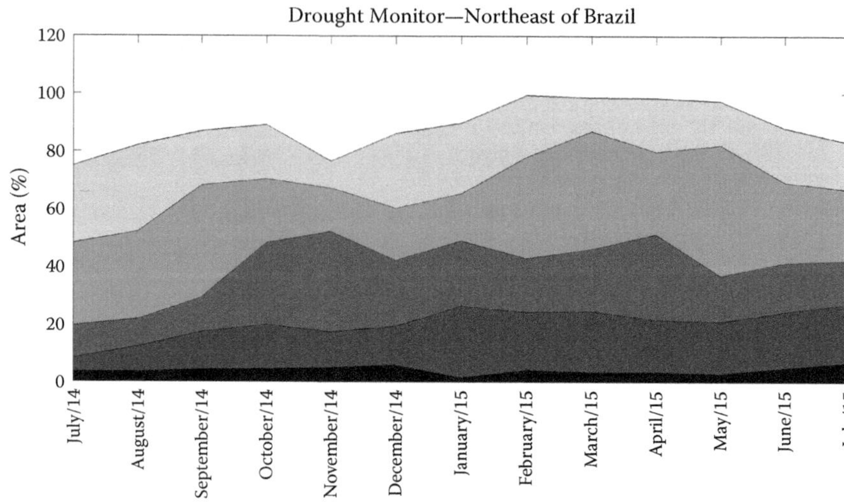

FIGURE 11.9
Evolution of the percentage of areas in different degrees of drought severity (D0–D4; D1–D4; D2–D4; D3–D4) from July 2014 to July 2015 for the Northeast region. They are also made available for each state.

The length of time from data preparation by the CI and publication of the Monitor is approximately eight days, which is important to maintain so that the results of the Drought Monitor can be published around the 15th day of the month, taking into account the time needed to compile the information, produce the map, and make current results available.

The publication process is in constant development, accompanying the updating and development of the Monitor and its support products which seek to attend to both technical users and the public in general. A restricted access site also exists for the authors, validators, CI, and the Technical Group that makes available the GIS projects and products for the monthly map design, the shapes of past Monitors, support material for the map design and validation, and other important documents in order to fully record the entire process.

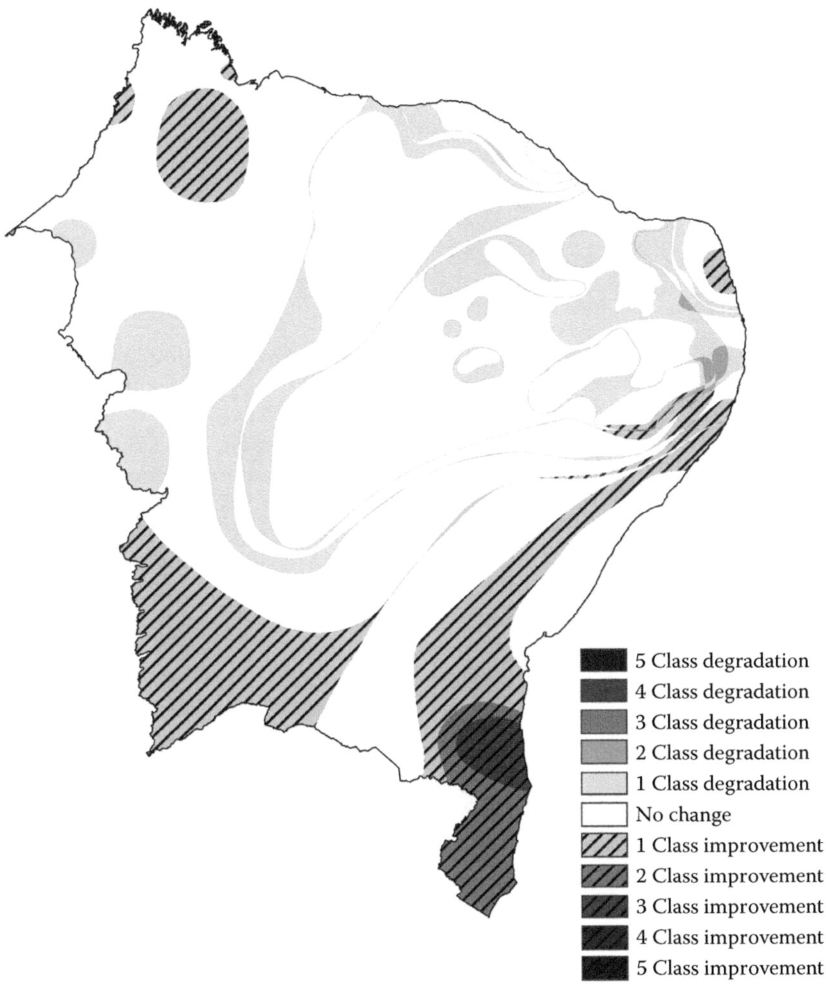

FIGURE 11.10
Map of the changes in drought severity categories from May to June 2015. Change maps will be made available for 1, 2, 3, 6, and 12 months, as well as for reference periods corresponding to the beginning of the calendar and hydrological years (January 1 and October 1, respectively).

12

Drought Preparedness Plans: Tools and Case Studies

Francisco de Assis de Souza Filho, Paula Pedreira de Freitas de Oliveira, Marcos Thadeu Abicalil, Cybelle Frazão Costa Braga, Samíria Maria Oliveira da Silva, Sandra Helena Silva de Aquino, Daniel Antonio Camelo Cid, Luiz Martins de Araújo Júnior, and Ana Cláudia Fernandes Medeiros Braga

CONTENTS

12.1 Introduction 168
12.2 Drought Indicators and Triggers 168
12.2.1 Context 168
12.2.2 Indicators and Performance Evaluation 169
12.2.2.1 Meteorological Indicators 169
12.2.2.2 Hydrological Indicators 170
12.2.2.3 Indicators Based on Storage Levels 171
12.2.2.4 Performance Evaluation 172
12.2.3 Application Example 174
12.3 Basin Drought Preparedness Plan 174
12.3.1 Context 174
12.3.2 Indicators and Drought Monitor for the Basin 179
12.3.2.1 Northeast Drought Monitor 180
12.3.2.2 Standardized Precipitation Index 182
12.3.2.3 Standardized Runoff Indicator 183
12.3.2.4 State Index 185
12.3.2.5 Drought Monitor for the Piranhas-Açu River Basin 186
12.3.3 Vulnerability Analysis 188
12.3.4 Actions 189
12.4 Drought Plan for a Water System 190
12.4.1 Context 190
12.4.2 Water System Operation 193
12.4.3 Actions 201
12.5 Urban Drought Preparedness Plan 202
12.5.1 Context 202
12.5.2 Monitoring and Early Warning 203

12.5.3 Vulnerability Analysis ... 203
12.5.4 Action Plan ... 206
12.5.5 Decision Support System ... 207
12.6 Final Observations ... 208

12.1 Introduction

Droughts and their impacts were widely discussed in the previous chapters of this book. As they are a recurring phenomenon and a part of climate variability, the development of drought preparedness plans (DPPs) constitute a useful tool to help reduce the economic, social, and environmental impacts associated with them.

The elaboration of a DPP requires the use of methodologies and tools that are appropriate for the system for which it is being developed. This chapter presents three case studies of drought preparedness planning for Northeast Brazil. The plans were developed for the states of Ceará, Rio Grande do Norte, Paraíba, and Pernambuco, more specifically for the Piranhas-Açu river basin (Paraíba and Rio Grande do Norte), the Jucazinho water system (Pernambuco), and the urban plan for the metropolitan region of Fortaleza (Ceará).

The chapter has five sections in addition to this one. The second section presents the indicators and triggers utilized in the DPPs. The third one presents the methodology for such plans at the river basin level, through the Piranhas-Açu river basin case study. The fourth section focuses on DPPs for water systems and presents the Jucazinho case study. The fifth one presents drought preparedness plans for urban areas, and the last section puts forward some final conclusions and recommendations.

12.2 Drought Indicators and Triggers

12.2.1 Context

Monitoring is an essential tool for operationalization of drought preparedness plans. Its purpose is to identify the evolution of drought severity over time through the use of indices designed to permit determination of different drought stages on the basis of predefined triggers. Such indices are extremely relevant for the elaboration of policies focused on the planning of mitigation actions for drought impacts (National Drought Policy Commission 2000). But the definition of drought indices and their thresholds is not a trivial task.

Normally, in order to detect a drought and evaluate its severity, indices based on variables such as temperature, rainfall, evaporation, evapotranspiration, soil humidity, water storage, and runoff are used. However, the complexity of this phenomenon makes it impossible for a single index to satisfactorily identify the various types of droughts, their severity, and their potential impacts. For each situation, the most appropriate index or set of indices should be used.

Various indices are described in the literature for the identification and monitoring of droughts in different timescales. Among these, the most disseminated include the Palmer drought severity index, PDSI (Keyantash and Dracup 2002), and the standardized precipitation index, SPI (McKee et al. 1993), the latter being associated with the standardized runoff index, SRI (Shukla and Wood 2008), and the Standardized Precipitation Evapotranspiration Index, SPEI (Vicente-Serrano et al. 2010).

Another index that is frequently used is the state index (SI). This index was used for the action plan for early warning and eventual occurrence of droughts in the Jucar and Ebro river basins in Spain in order to describe the droughts stage in the basins studied and to permit a quantitative comparison with other drought indices. In addition, another one that can be used as an indicator of the stage of a drought was developed by Cid et al. (2014) and refers to simulations for the proposal of operating rules for reservoirs, establishing target levels that can also serve as a drought index for a particular reservoir.

12.2.2 Indicators and Performance Evaluation

For a better understanding of drought indicators for surface water resource systems, meteorological drought indices, SPI and SPEI, and indices of hydrological droughts, SRI, SI and their equivalent index (IS, proposed in this chapter), together with the reservoir target level index, can be presented. A description of the method for the comparative or performance evaluation of these indices with their respective resolution follows.

12.2.2.1 Meteorological Indicators

The SPI is based on the distribution of rainfall probability and can be calculated for different timescales as it is a standardized index that permits comparison among different localities and climates. The SPEI considers the effects of precipitation and evapotranspiration jointly on drought severity and permits the identification of different types of droughts and their impacts on various systems.

For calculation of the SPI and the SPEI, the following steps are adopted: (1) select a timescale, such as, for example, total monthly precipitation or that for a period of several months (three months for instance); (2) adjust the probability distribution* for a time series of these values (frequently a gamma distribution); (3) estimate the probability values of not exceeding this distribution (i.e., the accumulated probability of precipitation values that are less

* A probability function is a function that assigns probabilities to the values of a random variable, here the n-months precipitation (n = 3, 4, 6, 12, 18, 24).

than this distribution in the time series) of precipitation for the year in which one wants to evaluate the drought; (4) calculate the reduced variable using a normal distribution; (5) seek the value corresponding to the probability of not exceeding the distribution. This procedure is detailed next.

The probability distribution utilized to adjust the data was the two-parameter gamma distribution defined by:

$$f(x) = \frac{1}{\beta^{\alpha}\Gamma(\alpha)} x^{\alpha-1} e^{-x/\beta}, \quad x > 0 \tag{12.1}$$

where:

$\alpha > 0$ is the parameter of form
$\beta > 0$ is the parameter of scale
x is the random variable under study
r is the gamma function

$$\Gamma(x) = \int_0^{\infty} y^{\alpha-1} e^{-y} dy \tag{12.2}$$

The gamma distribution, as indicated in Equation 12.1, is only defined for $x > 0$; thus, when a particular month presents a zero value, it is necessary to use a transformation of the accumulated probability distribution, given by:

$$(x) = q + (1-q)G(x) \tag{12.3}$$

in which $G(x)$, known as the *incomplete gamma function*, is the function of the distribution estimated from the nonzero values in the data series utilized, and q is the probability of zeros in the sample. With this the SPEI is defined, as is the SPI in Pereira and Paulo (2004) as:

$$\text{SPI} = \varphi^{-1}\left[F(x)\right] \tag{12.4}$$

in which φ is a function of the reduced normal distribution. The values of SPI and SPEI can be classified as indicated in Table 12.1.

12.2.2.2 Hydrological Indicators

For the calculation of SRI, river flow data are utilized and similar steps are followed to those used to calculate SPI. The flows represent synthetically the hydrological processes of the river basin associated with its cross-section. In monthly and seasonal timescales, the SRI is a useful complement to the SPI to depict hydrological aspects of the drought (Shukla and Wood 2008).

Thus, like the SPI and the SPEI, the SRI permits its application for different timescales. The SRI values can be classified as shown in Table 12.1.

TABLE 12.1
Thresholds of the Indices with Their Respective Classifications, Categories, and Stages

SPI, SPEI, and SRI	Classification	Category	Stage
>−0.79	D0	Abnormally dry	Pre-alert I
−0.80 to −1.29	D1	Moderate drought	Pre-alert II
−1.30 to −1.59	D2	Severe drought	Alert
−1.60 to −1.99	D3	Extreme drought	Emergency I
<−2.00	D4	Exceptional drought	Emergency II

Source: Cunha, R.L.A. Definição de cenários de referência para a avaliação dos impactos das secas, Masters' Thesis, Mestrado Integrado em Engenharia Civil -2007/2008—Departamento de Engenharia Civil, Faculdade de Engenharia da Universidade do Porto, Porto, Portugal, 2008.

12.2.2.3 Indicators Based on Storage Levels

The SI is an index that relates the state of a drought to the volume of stored water in a reservoir (Estrela et al. 2006). The index varies in a range from zero, corresponding to the minimum storage value for the reservoir, to one, corresponding to the maximum value of storage, being calculated according to the following equation:

$$\begin{aligned} &\text{If } V_i \geq V_{\text{med}} \Rightarrow I_e = \frac{1}{2}\left[1+\frac{V_i - V_{\text{med}}}{V_{\text{max}} - V_{\text{med}}}\right] \\ &\text{and if } V_i < V_{\text{med}} \Rightarrow I_e = \frac{V_i - V_{\text{min}}}{2(V_{\text{med}} - V_{\text{min}})} \end{aligned} \tag{12.5}$$

where:
V_i is the volume measured for the period analyzed
V_{med} is the average volume for the time period considered
V_{max} is the maximum volume
V_{min} is the minimum volume

For the IS case studies mentioned earlier, only the water volume values for December of each year in the time series analyzed were used. Given the observed quality of the data utilized and the short period of time for the accumulated water data available for this series, data simulated by a model of reservoir operation were also utilized, as proposed by Cid et al. (2014).

In the utilization of simulated water volume data by the model for reservoir operation, SI was relabeled as the synthetic index (IS). Complementarily, the SI as well as the IS can be calculated exchanging the average volume by the median volume, with the objective of reducing the effect of the size of the reservoir. The values of SI and IS are classified as indicated in Table 12.2.

TABLE 12.2

Categorization of the SI and IS Indices and the Target Levels

SI and IS	Target Levels	Stage
>0.5	D0	Pre-alert I
0.50–0.31	D1	Pre-alert II
0.30–0.16	D2	Alert
0.15–0.076	D3	Emergency I
<0.075	D4	Emergency II

Sources: Estrela, T. et al., Droughts and the European water framework directive: Implications on Spanish river basin districts, in: *Drought Management and Planning for Water Resources*, Andreu, J. et al. (ed.), CRC Press, Taylor & Francis Group, Boca Raton, FL, 2006, Chapter 6, pp. 169–191; Cid, D.A.C. et al., Uso de simulação para definição de níveis metas de operação para o reservatório Jucazinho/PE, in: XII Simpósio de Recursos Hídricos do Nordeste, 2014.

The average level of the reservoirs (described in Section 12.2.3) can serve as drought indices. In this case, each *zone* of the reservoir (i.e., reservoir region between two target levels) is associated with a drought stage. The values for the average levels associated with each drought stage are also presented in Table 12.2.

12.2.2.4 Performance Evaluation

Evaluation criteria for the capacity of the various indices to identify the occurrence of a given drought are desirable. Two common shortcomings of the indices are (1) a drought occurs but the index does not detect it (probability of detection [POD] of a drought); and (2) the drought does not occur but the index indicates that it has (false alarm).

In surface water resource systems with reservoirs, their levels would, theoretically, be the best drought indicators. However, there is an inconvenience with indicators based on the volumes stored due to the possibility of human error in reservoir operation. Given this possibility, the Synthetic Index for the reservoir, which represents its ideal operation, was created for this study. This indicator was considered to evaluate the capacity of the other indices (SPI, SPEI, and SRI) to predict the stage of the drought.

To evaluate the possible errors in predicting each stage of drought severity, the SI of volume was used as the reference index and the SPI, SPEI, and the SRI indices for the 12-, 24-, and 36-month timescales were evaluated for their capacity to predict the drought observed in the reservoir.

To undertake this evaluation, the contingency analysis method was used as a mathematical tool for comparison, and on this basis the POD of events, which is the ratio of the number of events that were correctly predicted and the total number of events, and the false alarm ratio (FAR), which is the proportion of predictions of drought events that, in fact, did not materialize, are shown in Equations 12.6 and 12.7, respectively (Schaefer 1990).

$$\text{POD} = \frac{a}{a+c} \tag{12.6}$$

$$\text{FAR} = \frac{b}{a+b} \tag{12.7}$$

with a being when the model predicted the event and it actually occurred, b when the model predicted the event and it did not occur, c when the model did not predict the event and it occurred, and d when the model did not predict the event and it did not occur. Table 12.3 shows a synthesis of the contingency analysis method.

For development of this quantitative analysis of drought indices, the case studies grouped the contingencies comparing the drought stages according to the SI and the target levels with the severity classification of SPI, SPEI, and SRI for the 12-, 24-, and 36-month timescales. This grouping was done according to the categories indicated in Table 12.4.

TABLE 12.3

Synthesis of the Contingency Analysis Method, with an Indication of the Calculations of POD and FAR

		Observation		
		Yes	**No**	**FAR**
Prediction	Yes	A	B	$b/(a+b)$
	No	C	D	
	POD	$a/(a+b)$		

Source: Amanajás, J.C., *Uso do método de contingência para análise da eficácia de previsão da precipitação pluviométrica do modelo ETA para o município de Macapá-AP em 2007*, Monograph, Federal University Amapá, Macapá, Brazil, 2008.

TABLE 12.4

Relation between the Groupings of the Stage and Degree of Severity of Droughts

	Degree of Drought Severity		
Drought Stage	**Humid and D0**	**D1 and D2**	**D3 and D4**
Normal	×		
Alert		×	
Emergency			×

Source: Elaboration by the authors.

12.2.3 Application Example

These indices and the performance evaluation will be shown in the case study for the Jucazinho water system, which is located in the Capibaribe river basin in the state of Pernambuco.

Table 12.5 shows the calculation of the annual time series for the different indices obtained for December of each year. It should be observed that the SPI and SPEI indices present drought detection problems in 1991, 1992, and 1993. The same occurred for SRI in 1971 and 1972. And a false alarm occurred for the storage system in 1981 and 1982.

This fact is associated with the multiyear regularization of the reservoir. This reservoir has a residence time (ratio between the maximum volume and the average annual discharge) of three years. If the reservoir was intra-annual in character (i.e., in which water from the rainy season is used for the dry season in the same year), possibly these detection failures would have been fewer.

Tables 12.6 and 12.7 present the calculated values of FAR and POD, considering a comparison between SI and the target levels with the SPI, SPEI, and SRI indices, respectively, for the three timescales for the Jucazinho water system.

In general, the POD values calculated reveal a low power of detection of the states of alert and emergency by SPI, SPEI, and SRI for all of the timescales considered in relation to the reference indices (SI and target levels). Among these, SRI12 and SPI24, with a POD value of just 0.11 for the state of emergency, in Table 12.6 and SPI36 and SRI36, with a POD value of 0.06 for the state of alert in Table 12.7 stand out in this regard. Of the values for FAR, one notes a greater general tendency for overestimation of the drought stage alert of the SPI, SPEI, and SRI indices. Thus, this suggests that the SPI, SPEI, and SRI indices are not good indicators of the drought stage when they refer to volumes of stored water.

12.3 Basin Drought Preparedness Plan

12.3.1 Context

Droughts and climate change generate strong stress on the environment, economy and society, which are exacerbated by the interconnections between them. Consequently, for sustainable water security, society needs to develop and strengthen its adaptive capacity (Thomas et al. 2013).

Drought management is intimately related to water resource management, as the latter recognizes that measures aiming at sustainable water use need to be adapted to climate conditions affecting any given river basin. It is also important to highlight that the planning unit for the Brazilian model of

TABLE 12.5

Year to Year Values of the SPI, SPEI, SRI, SI, IS, and Target-Level Indices for the 12-, 24-, and 36-Month Timescales, Related to the First Three Indices and the Classification of the Drought Stage, in which Yellow, Orange, Red, and Dark Red Represent the Stages of Pre-Alert, Alert, Emergency I, and Emergency II, Respectively

Year	Meteorological Drought						Hydrological Drought							
	Precipitation (P)			Balance (P-ETP)			Inflow (Q)			Stored Water (SW)				
	SPI-12	SPI-24	SPI-36	SPEI-12	SPEI-24	SPEI-36	SRI-12	SRI-24	SRI-36	IE_{median}	IE_{mean}	SI_{median}	SI_{mean}	Target Levels
1962	−0.68	–	–	−0.96	–	–	1.09	–	–	–	–	0.57	0.58	0
1963	−0.89	−1.20	–	−1.07	−1.33	–	−0.94	0.35	–	–	–	0.23	0.24	1
1964	2.49	1.39	0.82	2.14	1.06	0.19	0.65	−0.15	0.51	–	–	0.36	0.37	1
1965	−0.15	1.76	1.08	−0.28	1.65	0.62	0.09	0.33	−0.40	–	–	0.22	0.22	1
1966	1.42	0.88	2.31	1.40	0.79	2.33	1.37	1.03	1.15	–	–	0.66	0.67	0
1967	0.09	1.02	0.74	0.41	1.30	0.83	0.11	1.04	0.82	–	–	0.57	0.58	0
1968	−0.30	−0.24	0.66	0.07	0.18	1.05	−0.99	−0.78	0.44	–	–	0.22	0.23	1
1969	1.49	0.84	0.71	1.29	0.94	0.95	0.92	0.15	−0.10	–	–	0.50	0.51	1
1970	0.42	1.28	0.90	0.32	1.13	0.89	0.20	0.65	−0.03	–	–	0.40	0.41	2
1971	−0.84	−0.33	0.64	−0.85	−0.51	0.31	−0.11	−0.21	0.29	–	–	0.19	0.19	2
1972	0.05	−0.63	−0.35	0.42	−0.44	−0.28	−0.22	−0.55	−0.67	–	–	0.00	0.00	3
1973	0.22	0.09	−0.49	0.02	0.16	−0.44	−0.46	−0.83	−1.19	–	–	0.00	0.00	4
1974	1.39	1.08	0.90	1.31	0.93	0.96	1.17	0.58	0.17	–	–	0.55	0.55	1
1975	0.28	1.11	1.03	0.01	0.92	0.69	1.18	1.54	1.17	–	–	0.77	0.77	1
1976	−1.12	−0.62	0.35	−1.06	−0.83	0.00	−1.61	0.36	0.97	–	–	0.50	0.51	0
1977	2.01	0.86	0.83	1.85	0.69	0.48	1.74	1.02	1.55	–	–	0.96	0.96	0
1978	1.32	2.26	1.48	1.25	2.44	1.36	0.71	1.72	1.19	–	–	1.00	1.00	0
1979	−0.85	0.39	1.59	−0.60	0.44	1.67	−1.63	−0.22	1.19	–	–	0.72	0.73	0
1980	−0.99	−1.40	−0.28	−0.80	−1.00	−0.23	−0.56	−1.84	−0.89	–	–	0.53	0.54	0
1981	−1.63	−1.93	−2.28	−1.18	−1.30	−1.37	1.34	0.74	−0.03	–	–	0.80	0.81	0

(Continued)

TABLE 12.5 (*Continued*)

Year to Year Values of the SPI, SPEI, SRI, SI, IS, and Target-Level Indices for the 12-, 24-, and 36-Month Timescales, Related to the First Three Indices and the Classification of the Drought Stage, in which Yellow, Orange, Red, and Dark Red Represent the Stages of Pre-Alert, Alert, Emergency I, and Emergency II, Respectively

Year	Meterological Drought						Hydrological Drought							
	Precipitation (P)			Balance (P-ETP)			Inflow (Q)			Stored Water (SW)				
	SPI-12	SPI-24	SPI-36	SPEI-12	SPEI-24	SPEI-36	SRI-12	SRI-24	SRI-36	IE_{median}	IE_{mean}	SI_{median}	SI_{mean}	Target Levels
1982	−0.47	−1.52	−2.02	−0.45	−1.13	−1.31	−0.24	0.85	0.33	–	–	0.65	0.65	0
1983	−0.90	−1.05	−1.96	−0.91	−0.98	−1.35	−1.71	−1.44	0.10	–	–	0.29	0.30	2
1984	0.97	0.07	−0.30	0.99	−0.06	−0.38	0.63	−0.33	−0.81	–	–	0.40	0.41	2
1985	1.11	1.36	0.69	1.08	1.46	0.56	1.49	1.44	0.84	–	–	0.79	0.79	0
1986	−0.04	0.70	1.11	0.33	0.94	1.36	0.80	1.54	1.66	–	–	0.87	0.87	0
1987	0.08	−0.07	0.57	0.54	0.48	1.04	−0.61	0.13	1.13	–	–	0.66	0.66	0
1988	0.08	0.01	−0.11	0.58	0.67	0.67	0.44	−0.26	0.13	–	–	0.64	0.65	0
1989	0.08	0.01	−0.03	0.54	0.67	0.82	0.76	0.66	0.09	–	–	0.72	0.72	0
1990	0.08	0.01	−0.03	0.54	0.65	0.82	−0.37	0.19	0.18	–	–	0.55	0.55	0
1991	0.08	0.01	−0.03	0.54	0.65	0.79	−0.79	−1.20	−0.52	–	–	0.21	0.22	2
1992	0.08	0.01	−0.03	0.58	0.67	0.82	−0.87	−1.66	−2.13	–	–	0.00	0.00	4
1993	−0.73	−0.53	−0.50	−0.28	0.06	0.29	−1.71	−2.33	−3.10	–	–	0.00	0.00	4
1994	0.80	0.04	−0.01	0.74	0.19	0.42	1.19	0.35	−0.35	–	–	0.57	0.58	3
1995	−0.53	0.16	−0.37	−0.63	−0.06	−0.29	0.18	0.89	0.16	–	–	0.51	0.51	2
1996	−0.59	−0.88	−0.29	−0.64	−0.93	−0.48	0.45	0.22	0.88	–	–	0.51	0.51	1
1997	0.04	−0.47	−0.82	0.01	−0.57	−0.84	−0.19	−0.03	−0.21	–	–	0.27	0.28	2
1998	−2.52	−1.57	−1.81	−2.50	−1.67	−1.65	−1.21	−1.22	−0.91	–	–	0.00	0.00	4
1999	−1.64	−2.99	−2.45	−1.87	−2.31	−2.01	−1.00	−2.16	−2.23	–	–	0.00	0.00	4

(*Continued*)

TABLE 12.5 (*Continued*)

Year to Year Values of the SPI, SPEI, SRI, SI, IS, and Target-Level Indices for the 12-, 24-, and 36-Month Timescales, Related to the First Three Indices and the Classification of the Drought Stage, in which Yellow, Orange, Red, and Dark Red Represent the States of Pre-Alert, Alert, Emergency I, and Emergency II, Respectively

Year	Meterological Drought						Hydrological Drought							
	Precipitation (P)			Balance (P-ETP)			Inflow (Q)			Stored Water (SW)				
	SPI-12	SPI-24	SPI-36	SPEI-12	SPEI-24	SPEI-36	SRI-12	SRI-24	SRI-36	IE_{median}	IE_{mean}	SI_{median}	SI_{mean}	Target Levels
2000	1.75	0.38	−0.94	1.54	−0.32	−1.54	0.42	−0.43	−1.40	0.13	0.15	0.24	0.25	4
2001	−0.50	0.95	−0.04	−0.63	0.68	−0.67	0.48	0.42	−0.31	0.11	0.13	0.28	0.29	3
2002	0.00	−0.44	0.76	−0.35	−0.78	0.25	−0.09	0.06	0.06	0.08	0.10	0.09	0.09	3
2003	−0.93	−0.72	−1.03	−1.37	−1.19	−1.30	−1.40	−1.17	−0.84	0.00	0.00	0.00	0.00	4
2004	1.35	0.38	0.24	1.19	−0.24	−0.47	1.23	0.43	0.07	0.89	0.92	0.51	0.51	1
2005	0.35	1.13	0.45	−0.09	0.73	−0.33	0.84	1.33	0.74	1.00	1.00	0.61	0.62	0
2006	−0.79	−0.36	0.51	−1.22	−0.97	−0.23	−0.36	0.28	0.96	0.68	0.76	0.41	0.42	1
2007	−0.48	−0.99	−0.70	−0.90	−1.37	−1.24	−0.79	−1.18	−0.42	0.42	0.50	0.05	0.05	3
2008	−0.04	−0.46	−0.97	−0.32	−0.92	−1.32	–	–	–	0.93	0.95	0.00	0.00	–
2009	−0.60	−0.54	−0.85	−1.11	−1.04	−1.28	–	–	–	0.94	0.95	–	–	–
2010	0.38	−0.21	−0.30	−0.29	−1.02	−1.05	–	–	–	0.94	0.96	–	–	–
2011	1.17	1.01	−0.50	0.85	0.27	−0.50	–	–	–	0.95	0.96	–	–	–
2012	−1.37	−0.01	0.14	–	–	–	–	–	–	0.33	0.39	–	–	–

TABLE 12.6

Values of FAR and POD Resulting from the Comparison between SI and SPI, SPEI, and SRI Indices for the 12-, 24-, and 36-Month Timescales for the Jucazinho Water System

		SPI12		SPI24		SPI36	
		FAR	POD	FAR	POD	FAR	POD
Synthetic index (average)	Normal	0.47	0.83	0.50	0.82	0.44	0.91
	Alert	0.57	0.21	0.71	0.14	0.33	0.15
	Emergency	0.33	0.22	0.50	0.11	0.60	0.22
		SPEI12		SPEI24		SPEI36	
		FAR	POD	FAR	POD	FAR	POD
	Normal	0.43	0.87	0.48	0.77	0.43	0.91
	Alert	0.56	0.29	0.70	0.21	0.57	0.23
	Emergency	0.00	0.22	0.00	0.22	0.00	0.22
		SRI12		SRI24		SRI36	
		FAR	POD	FAR	POD	FAR	POD
	Normal	0.42	0.91	0.40	0.95	0.40	0.95
	Alert	0.67	0.14	0.67	0.14	0.67	0.15
	Emergency	0.75	0.11	0.25	0.33	0.00	0.33

Source: Elaboration by the authors.

water resource management is the river basin (Brazil Water Law 1997) and that this scale should also be adopted for drought management (Figure 12.1).

The experience of the Piranhas-Açu river basin is described in this context. This basin was the object of the elaboration of a Drought Protocol* in order to construct a planning instrument for better coexistence with the drought and with a focus on preparedness for (and not only responses to) a drought period.

The Piranhas-Açu river basin is located in the semiarid region of Northeast Brazil (Figure 12.2). It possesses a total drainage area of 43,681.5 km^2, corresponding to 60% of the area of the state of Paraíba and 40% of that of Rio Grande do Norte. Thus, its waters come under both state and federal jurisdiction.† The basin includes 147 municipalities and has an estimated population of 1,398,820 inhabitants.

* For the Piranhas-Açu basin, it was decided to use the name *Drought Preparedness Protocol* to avoid confusion with the Basin Plan. However, the Protocol was incorporated in the action program within the Basin Plan.

† The Brazilian Constitution establishes a distinction between federally controlled water, for rivers across state boundaries, and state-controlled water, for rivers and groundwater that remain completely within state boundaries.

TABLE 12.7

Values of FAR and POD Resulting from the Comparison between Target Levels and the SPI, SPEI, and SRI Indices for the 12-, 24-, and 36-Month Timescales for the Jucazinho Water System

		SPI12		SPI24		SPI36	
		FAR	POD	FAR	POD	FAR	POD
Target levels	Normal	0.64	0.76	0.64	0.81	0.61	0.88
	Alert	0.57	0.18	0.57	0.18	0.67	0.06
	Emergency	0.33	0.17	0.50	0.08	0.60	0.17
		SPEI12		SPEI24		SPEI36	
		FAR	POD	FAR	POD	FAR	POD
	Normal	0.60	0.82	0.64	0.75	0.60	0.88
	Alert	0.56	0.24	0.60	0.24	0.71	0.13
	Emergency	0.00	0.17	0.00	0.17	0.00	0.17
		SRI12		SRI24		SRI36	
		FAR	POD	FAR	POD	FAR	POD
	Normal	0.58	0.88	0.57	0.94	0.57	0.94
	Alert	0.67	0.12	0.67	0.12	0.83	0.06
	Emergency	0.75	0.08	0.25	0.25	0.00	0.25

Source: Elaboration by the authors.

Among its principal characteristics are the intermittent rivers, high spatial and temporal variability of precipitation, and rainfall concentrated in a few months, generally between February and May, a period in which surface discharges and recurrent droughts also occur. The main water storage is through reservoirs, both annual and multiyear regulatory ones, which are spatially distributed across the basin. (Agência Nacional de Águas 2014).

For elaboration of the Drought Preparedness Protocol for the Piranhas-Açu River Basin, an approach focused on the water availability question and anchored in the pillars proposed by Wilhite et al. (2005) was adopted. Figure 12.3 presents a flowchart of the activities that were developed.

The following sections detail the monitoring (i.e., indicators adopted), vulnerability analysis, and proposed actions that composed the Protocol for Drought Preparedness for the Piranhas-Açu River Basin.

12.3.2 Indicators and Drought Monitor for the Basin

The Drought Protocol for the Piranhas-Açu River Basin was based on indicators capable of monitoring the drought categories in the basin and triggering the necessary actions to prepare for, adapt to, and mitigate their effects. For this, it was important that the indicators could be easily interpreted and

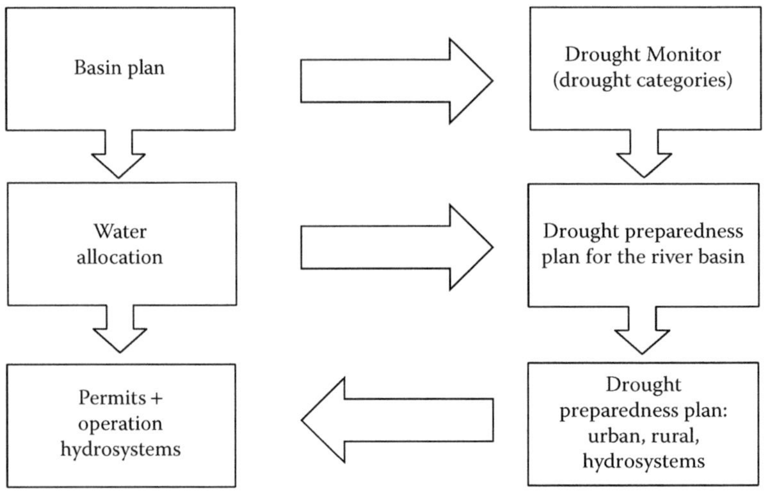

FIGURE 12.1
Relation between the Basin Plan and the Drought Preparedness Protocol for the Piranhas-Açu River Basin.

representative of the characteristics of the region, including the different Hydrological Planning Units (HPUs)* for the basin.

The indicators for the basin Drought Monitor were selected on the basis of both a regional (Northeast Drought Monitor and SPI) and local (SRI and SI) perspective.

12.3.2.1 Northeast Drought Monitor

The information and technical knowledge about the droughts that occur in Northeast Brazil, until recently, were dispersed among various government agencies. The need for a better understanding of the droughts, including their severity, their spatial and temporal evolution, and their impacts, resulted in construction a Northeast Drought Monitor. As described more fully in Chapter 4 and 11, the Monitor constitutes a process of regular accompaniment of the droughts in the region.† Each month, the Monitor process results in the production of a map that reflects the physical or natural drought, based on meteorological, hydrological and agricultural data, and depicts its severity through five drought stages or categories (D0 to D4).

* A HPU is a subdivision of the basin established in the Basin Plan for the Piranhas-Açu River (Agência Nacional de Águas 2014).

† monitordesecas.ana.gov.br.

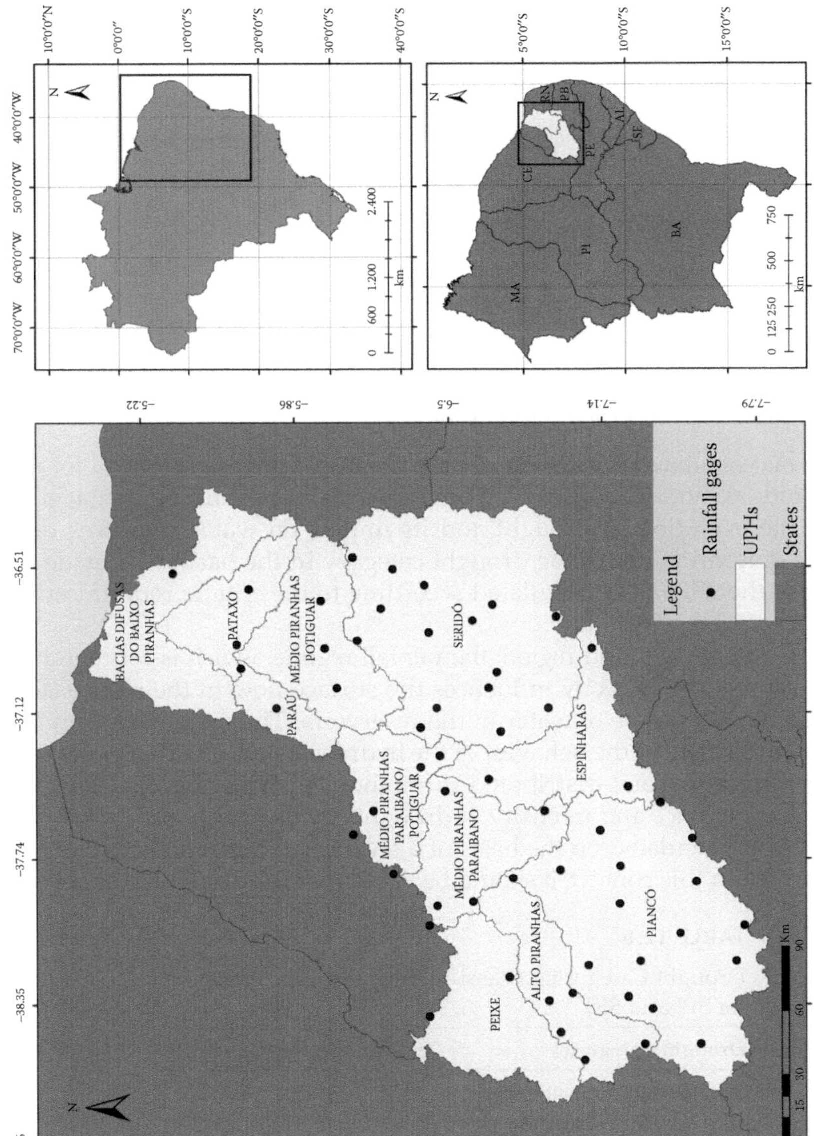

FIGURE 12.2
River basin of the Piranhas-Açu river.

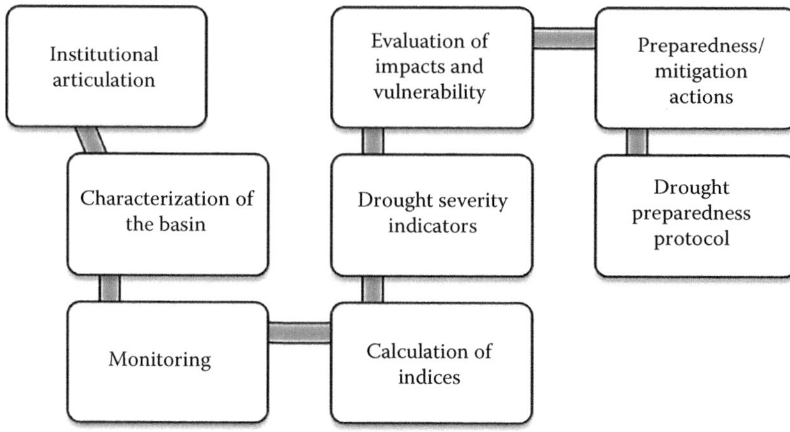

FIGURE 12.3
Activities of Drought Preparedness Plan elaboration for the Piranhas-Açu River Basin.

12.3.2.2 Standardized Precipitation Index

The SPI was calculated with average precipitation data for each HPU for the 12-, 24-, and 36-month timescales. These timescales permit the manager to evaluate the evolution of drought and its impact on water resources over time. The determination of the drought category in the basin was made on the basis of the SPI values calculated according to the criteria reproduced in Table 12.8.

In the Northeast semiarid region, the rainfall regime, which is concentrated during four months, directly influences the surface flow in the river basins and hence of the volume of water in the reservoirs. Thus, the long-term SPI values are indicative of the behavior of the hydrosystems, and especially that of the smaller reservoirs distributed throughout the river basin. They also reflect the occurrence and intensity of drought events over time.

The SPI was calculated on the basis of a data series between January 1962 and June 2014. In this context, it should be noted that the drought experienced

TABLE 12.8
Drought Categories, Classification, and Thresholds for SPI and SRI

Drought Categories		Trigger—SPI and SRI
D0	Abnormally dry	−0.5 to −0.79
D1	Moderate drought	−0.8 to −1.29
D2	Severe drought	−1.3 to −1.59
D3	Extreme drought	−1.6 to 1.99
D4	Exceptional drought	<−2.0

Source: Svoboda, M. et al., The drought monitor, *Bull. Am. Meteorol. Soc.*, 83(8), 1181, 2002.

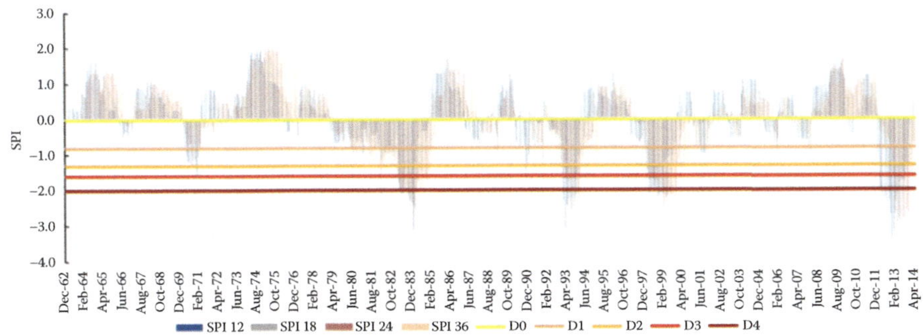

FIGURE 12.4
SPIs for the Seridó HPU.

between 2012 and 2014 was the one of greatest intensity over the period analyzed in the majority of the HPUs. In general, the most severe droughts (i.e., categories D3 and D4) occurred between 1980 and 1983, 1993–1994, 1998–1999, and 2012–2014, which impacted all the HPUs, although with some small time lags and different intensities. Other less intense droughts also occurred in other time periods. Figure 12.4 exemplifies this situation by presenting the temporal distribution of the SPI values for the Seridó HPU.

Table 12.9 presents the SPIs for 12, 18, 24, and 36 months calculated for the HPUs of Peixe and Seridó between June 2012 and May 2013 and the volumes of water stored in the Lagoa do Arroz and Itans reservoirs. The most severe drought categories are consistent with the drop in water volumes in the reservoirs, with the SPI values correctly reflecting the degree of criticality of the droughts.

In general, the SPI for 12 months (SPI12) was the first that indicated the most severe drought category. However, in some HPUs, the change to the most critical category occurred (evaluating the SPI values over all three timescales) without full recovery of the water volumes stored in the reservoirs. In these cases, the rainfall that occurred in the region was not sufficient to recharge the reservoirs, but contributed enough for the drought category to become less severe. This aspect reinforces the importance of evaluating more than one indicator to determine the drought category in the basin.

12.3.2.3 Standardized Runoff Indicator

The SRI was calculated on the basis of synthetic incremental tributary discharges* to the 52 strategic reservoirs† in the basin for the six-month timescale. The SRI has the same calculation procedures and triggers as the SPI (see Table 12.8).

* The discharges were calculated with a rainfall-runoff model called SMAP.
† Established in the Basin Plan (Agência Nacional de Águas 2014).

TABLE 12.9

SPI Values for the Peixe and Seridó HPUs and Percentage Water Volume of the Lagoa do Arroz and Itans Reservoirs

Ano	UPH Peixe						UPH Seridó					
	SPI12	SPI18	SPI24	SPI36	Category	Res. lagoa do arroz volume (%)	SPI12	SPI18	SPI24	SPI36	Category	Res. itans volume (%)
June-12	−0.92	−0.36	−0.25	−0.61	D0	49.48	−1.96	−0.42	−0.24	−0.38	D3	53.50
July-12	−1.04	−0.96	−0.24	−0.7	D1	46.52	−2.24	−0.63	−0.24	−0.49	D4	50.58
August-12	−1.04	−1.36	−0.24	−0.76	D1	41.49	−2.26	−0.75	−0.24	−0.57	D4	47.49
September-12	−1.04	−0.97	−0.24	−0.76	D1	37.06	−2.25	−0.77	−0.24	−0.58	D4	44.88
October-12	−1.57	−1.07	−0.4	−0.78	D2	31.02	−2.31	−1.33	−0.52	−0.58	D4	41.92
November-12	−1.74	−1.11	−0.39	−0.77	D3	27.80	−2.36	−2.1	−0.52	−0.58	D4	37.42
December-12	−1.72	−1.21	−0.58	−0.93	D3	24.73	−2.31	−2.12	−0.62	−0.66	D4	33.83
January-13	−2	−1.68	−1.38	−1.08	D4	21.92	−2.69	−2.58	−0.92	−0.87	D4	−99.00
February-13	−1.99	−1.6	−1.78	−0.94	D3	20.14	−3.38	−2.84	−1.31	−1.01	D4	28.18
March-13	−1.98	−1.83	−1.59	−0.81	D3	20.39	−3.19	−3.1	−1.63	−1.04	D4	26.92
April-13	−1.96	−2.02	−1.61	−1.06	D3	19.60	−2.67	−3.12	−2.35	−1.11	D4	25.28
May-13	−1.52	−1.98	−1.52	−0.85	D3	19.20	−2.34	−3.14	−2.95	−1.12	D4	23.77

Note: −99.00 indicates a false number; color scale in accordance with Table 12.5.

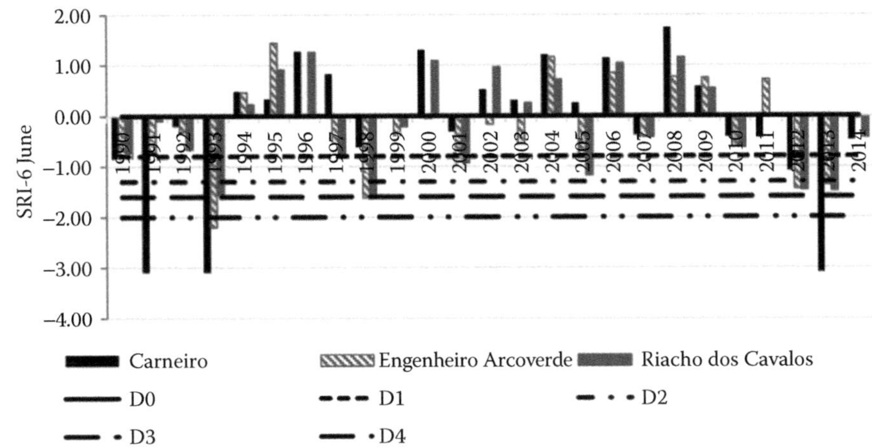

FIGURE 12.5
SRI-6 in the month of June for the HPU Médio Piranhas Paraibano between 1990 and 2014.

In general, the SRI indicated the drought periods in the basin as a whole. The most intense droughts influenced the inflow to the majority of the reservoirs, as in 1970, 1982, 1983, 1993, 1998, 1999, 2012, and 2013. In other drought years, this effect was not uniform in the same HPU and the SRI indicated areas without a drought category or with different categories, as can be observed in Figure 12.5 for 1991 and 1997 in the HPU of Médio Piranhas Paraibano (Carneiro, Engenheiro Arcoverde, and Riacho dos Cavalos reservoirs). Thus, the reservoir discharges can result in different drought categories for the same HPU.

12.3.2.4 State Index

The State Index (SI) is directly related to the historical and present volumes of water in the reservoir and should have been calculated for the 52 strategic reservoirs in the basin. However, there was a problem with the quality of the time series of the volumes of water stored in the reservoirs and it was decided to calculate the SI for the reservoirs that met the following conditions: (1) maximum 30% of monthly defects; and (2) a minumum of 15 years in the time series. As the monitoring of the reservoirs and the length and quality of the time series improve, more reservoirs can be considered in the Drought Monitor for the basin.

The limits of this index were adapted from the categories of the Drought Preparedness Plan for Jucazinho (PES 2007) for the present case, since with these thresholds, the most severe categories were triggered late, with the reservoir practically dry. In this sense, the data were calibrated and new thresholds were adopted, as indicated in Table 12.10.

Figure 12.6 presents the temporal distribution of SI for the Coremas-Mãe d'água reservoirs in the Piancó HPU. The most intense droughts occurred

TABLE 12.10

Drought Categories, Classification, and the Respective Thresholds for SI for the Piranhas-Açu River Basin

Drought Category		Trigger—SI
D0	Abnormally dry	>0.6
D1	Moderate drought	0.4–0.6
D2	Severe drought	0.3–0.4
D3	Extreme drought	0.15–0.3
D4	Exceptional drought	<0.15

Source: Estrela, T. et al., Droughts and the European water framework directive: Implications on Spanish river basin districts, in: *Drought Management and Planning for Water Resources*, Andreu, J. et al. (ed.), CRC Press, Taylor & Francis Group, Boca Raton, FL, 2006, Chapter 6, pp. 169–191.

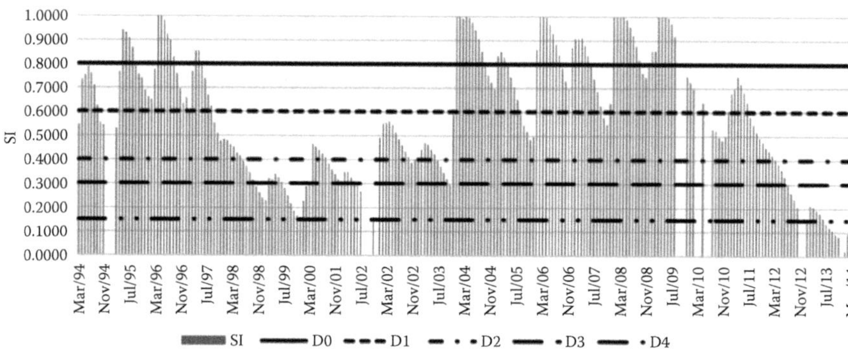

FIGURE 12.6
State Index (SI) for the Coremas-Mãe d'água in the Piancó HPU.

between 1998–1999 and 2012–2014, with the latter one being the most severe in terms of the volumes of water stored in the reservoir during the period studied.

The drought is a process that occurs slowly, and one can observe a natural progression in terms of the change of drought stage from less to more severe. However, the inverse situation does not always occur. In the analysis of the SI values calculated for the basin, some reservoirs were identified that received an elevated inflow of water in a short period of time and, as a result, experienced a *jump* in drought category, passing, for example, directly from D2 to D0.

12.3.2.5 Drought Monitor for the Piranhas-Açu River Basin

The Drought Monitor for the Piranhas-Açu River Basin (DM-PARB) was proposed to respond to the need for monitoring on a scale sufficient to define the drought categories for the basin and to trigger the actions necessary to mitigate the effects of the drought. This was needed because the area that the Northeast Drought Monitor covers is 1,554,291.74 km²—i.e., much larger than the 43,676.47 km^2 covered by the Piranhas-Açu (PA) basin, which is subdivided into a number of HPUs, among which there are important hydrological parameters that affect drought severity such as surface discharges and water storage in the reservoirs.

As for the Northeast Drought Monitor, monitoring in the PA basin is continuous, with monthly updates. Information is released each 15th day of the month, based on indicators calculated with data collected during the preceding 30 days. After calculation of the indicators and definition of the drought categories, a joint visualization of the indicators is done through a Table (see Table 12.11, with just two reservoirs in each HPU) and a technical note

TABLE 12.11

Drought Monitor for the Piranhas-Açu River Basin, June 2014

	Indicators/Categories				
	Regional (by HPU)		Local		
HPU	Northeast Drought Monitor	SPI	Reservoirs	SRI	SI
Bacias Difusas do Baixo Piranhas	D1	D4	Boqueirão de Angicos	Without relative drought (WRD)	*
Pataxó	D0	D2	Pataxós	WRD	*
Parau	D1	D0	Mendubim	WRD	*
			Beldroega	WRD	*
Médio Piranhas Potiguar	D0	D3	Armando Ribeiro Goncalves	D0	4
Seridó	D0	D3	Boqueirão de Parelhas	WRD	*
			Itans	WRD	*
Médio Piranhas Paraibano/Potiguar	D0	D0	Baião	D4	*
			Tapera	WRD	*
Médio Piranhas Paraibano	D0	D0	Carneiro	WRD	**
			Engenheiro Arcoverde	WRD	D4
Espinharas	D0	D0	Farinha	WRD	D2
			Jatobá I	WRD	D3
Peixe	D2	D0	Lagoa do Arroz	WRD	D3
			Capivara	WRD	**
Alto Piranhas	D1	D1	Engenheiro Ávidas	D4	D4
			São Gonçalo	D0	D3
Piancó	D1	D3	Coremas-Mãe d'água	WRD	D4
			Saco	WRD	**

Note: * predominant category; ** reservoirs with insuffucient data to calculate SI.
This way, the DM-PARB, togther with the vulnerability analysis, permits the selection of actions and facilitates the decision making process for the choice of drought preparedness and response measures.

that summarizes the analyses undertaken on the basis of the information collected regarding the severity, evolution, and impacts of the drought in the basin. It should be emphasized, however, that new indicators can and should be incorporated, but always in a way consistent with the evolution of the Monitor.

12.3.3 Vulnerability Analysis

The vulnerability analysis can also contribute to orient decision makers in the adoption of drought preparedness actions since it indicates the aspects and locations where greatest support is needed. It is important to assess what parts of the basin are most vulnerable and to generate theoretical and conceptual inputs for the effective management of these risks. Various authors have proposed indicators and methodologies in relation to drought and vulnerability (Bhattacharya and Dass 2007; Salvati et al. 2009; Antwi-Agyei 2012).

In elaborating the Drought Protocol, it was decided to use the methodology developed by Bhattacharya and Dass as adapted by Rosendo (2014) for the Brazilian semiarid region, which takes into account its specificities together with national policies for coexistence with the drought. Vulnerability is expressed by the relation (1 = arithmetic mean) between the indicators of exposure, sensitivity, and adaptive capacity (see Figure 12.7).

Table 12.12 presents the results of the vulnerability analysis for the PA basin by HPU. It is important to point out that a value obtained that is near 1 for the indicators for exposure and sensitivity shows the areas that are most exposed and/or sensitive to the drought event, thus representing a negative factor. In relation to the capacity indicator, on the other hand, the opposite occurs, since the greater the ability to adapt and, thereby, suffer less harm as a result of the drought—or even when damage is experienced, when it is less severe—the better.

The results of the analysis indicate that the Seridó HPU presents one of the lowest levels of vulnerability as a result of its comparatively greater adaptive

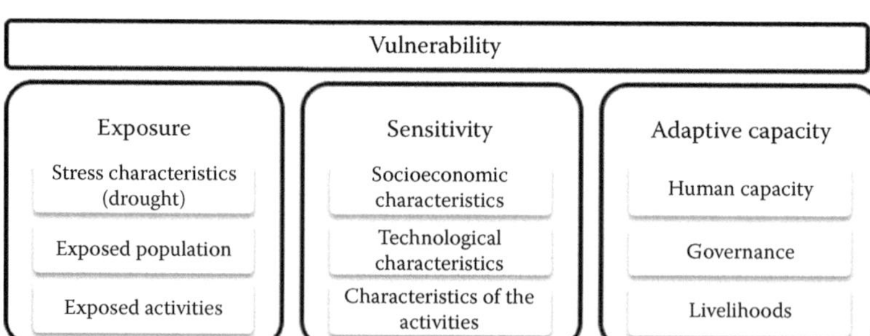

FIGURE 12.7
Indicators of drought vulnerability.

TABLE 12.12

Indicators of Exposure, Sensitivity, Adaptive Capacity, and Vulnerability for the Piranhas-Açu River Basin

HPU	Exposure	Sensitivity	Adaptive Capacity	Vulnerability
Bacias Difusas do Baixo Piranhas	0.420	0.387	0.510	0.561
Pataxó	0.337	0.459	0.371	0.611
Parau	0.360	0.456	0.315	0.623
Médio Piranhas Potiguar	0.386	0.396	0.366	0.617
Médio Pirahhas Paraibano-Potiguar	0.332	0.507	0.322	0.613
Seridó	0.548	0.463	0.402	0.529
Peixe	0.266	0.514	0.396	0.608
Médio Piranhas Paraibano	0.319	0.530	0.340	0.640
Espinharas	0.275	0.435	0.388	0.634
Alto Piranhas	0.242	0.511	0.115	0.711
Piancó	0.303	0.462	0.175	0.687

capacity despite its strong exposure and sensitivity. This reinforces the importance of preparing subregions to confront the drought in advance and not just to focus on response actions. It should also be noted that the upstream areas of the basin (HPU Piancó and HPU Alto Piranhas) are the most vulnerable and, thus, should receive the greatest attention from the water resources managers.

12.3.4 Actions

The combination of drought category and the evaluation of impacts and vulnerability indicates the *urgency* of preparedness and mitigation for different drought scenarios and, thus, the need for the implementation of strategic, tactical, or emergency actions (Figure 12.8 and Table 12.13). The drought monitoring indicators for the basin are considered to be triggers for the definition of different drought categories and, in consequence, the degree of severity of the drought (Table 12.13).

The actions proposed in the Drought Preparedness Protocol for the Piranhas-Açu River Basin were grouped together in thematic axes related to the drought. Table 12.14 presents some of these actions. It is important that water resources managers review the actions in the drought categories

FIGURE 12.8
Preparation for droughts.

TABLE 12.13

Protocol for Drought Preparedness in the Piranhas-Açu River Basin

		D0	D1	D2	D3	D4
Drought Category		**Abnormally Dry**	**Moderate Drought**	**Severe Drought**	**Extreme Drought**	**Exceptional Drought**
Triggers	SPI	−0.5 to −0.79	−0.8 to −1.29	−1.3 to −1.59	−1.6 to −1.99	−2.0
	SRI	−0.5 to −0.79	−0.8 to −1.29	−1.3 to −1.59	−1.6 to −1.99	−2.0
	SI	>0.6	0.4–0.6	0.3–0.4	0.15–0.3	>0.15
State		Pre-alert I	Pre-alert II	Alert	Emergency I	Emergency II
Type of action		Strategic	Strategic	Tactical	Emergency	Emergency

subsequent to the one that applies at any given point in time. However, it is also important to recall that the Protocol was incorporated into the Basin Plan through specific water resource management actions associated with preparing for and/or responding to the drought.

Drought preparedness planning is a process and not a discrete event. After the Drought Preparedness Protocol is elaborated, its implementation should be accompanied and the measures proposed be reviewed in order to maintain it up to date in terms of technological, legal, institutional, political, and management changes, as well as the evolving needs of the basin.

For purposes of implementation, moreover, strong institutional articulation within the basin is also important to ensure the *ownership* of the Protocol by regional stakeholders, establishing a favorable and collaborative scenario for its execution.

Review of the Protocol should focus on the evaluation of its effectiveness, especially of the monitoring and the actions taken, as well as the performance of those responsible, including weak points and shortcomings. The frequency of this review should be determined by the key stakeholders involved, as the process matures over time. At the time of each review, it is recommended to establish a new Protocol.

12.4 Drought Plan for a Water System

12.4.1 Context

The characterization of the drought stage in a water system is fundamental for the definition of measures that can effectively mitigate and respond to the damages caused by water scarcity. Accordingly, this book proposes the

TABLE 12.14

Drought Management Actions for the Piranhas-Açu River Basin

	Actions/Description				
Thematic Axis	**D0—Pre-Alert I**	**D1—Pre-Alert II**	**D2—Alert**	**D3—Emergency I**	**D4—Emergency II**
Communication	Information to the public through the media, highlighting the drought category and state	Information to the public through the media, highlighting the drought category and state	Information to the public through the media, highlighting the drought category and state	Information to water users, those registered and holding permits about use restrictions established	Information to water users, those registered and holding permits about use restrictions established
	Diagnosis of water systems and demand met—bimonthly updates	Educational campaigns for rational water use adjusted to the drought category	Educational campaigns for rational water use adjusted to the drought category	Information to the public through the media, highlighting the drought category and state	Information to the public through the media, highlighting the drought category and state
	Educational campaigns for rational water use adjusted to the drought category	Diagnosis of water systems and demand met—bimonthly updates	Communication to water users holding permits	Educational campaigns for rational water use adjusted to the drought category	Educational campaigns for rational water use adjusted to the drought category
Operation of water systems	Operation of water systems in accordance with alert levels and drought categories	Operation of water systems in accordance with alert levels and drought categories	Operation of water systems in accordance with daily alert levels and drought categories	Operation of water systems in accordance with daily alert levels and drought categories	Operation of water systems in accordance with daily alert levels and drought categories
	Use of decision support systems (DSSs)	Semiannual evaluation of support capacity of the reservoirs and water quality	Joint operation of reservoirs with use of DSSs and climate forecasts	Joint operation of reservoirs with use of DSSs and climate forecasts	Joint operation of reservoirs with use of DSSs and climate forecasts

(Continued)

TABLE 12.14 (*Continued*)

Drought Management Actions for the Piranhas-Açu River Basin

	Actions/Description				
Thematic Axis	**D0—Pre-Alert I**	**D1—Pre-Alert II**	**D2—Alert**	**D3—Emergency I**	**D4—Emergency II**
Urban water use (household and industrial)	Loss reduction—water supply systems	Loss reduction—water supply systems	Loss reduction—water supply systems	Loss reduction—water supply systems	Loss reduction—water supply systems
	Increased metering	Reduction of consumption x% by water supply system consumers	Reduction of consumption x% by water supply system consumers	Reduction of consumption y% by water supply system consumers	Reduction of consumption y% by water supply system consumers
	Reduction of consumption by water supply system consumers	Diagnostic of system operation	Evaluation/ adjustments in system operation	Adjustments in the network/rationing	Adjustments in the network/rationing
	Expansion of water supply and sewage systems in response to demand	Evaluation of alternative water supply sources	Evaluation of alternative water supply sources	Use of alternative sources of water supply (tank trucks, rapid construction aqueducts)	Use of alternative sources of water supply (tank trucks, rapid construction aqueducts)

Source: World Bank (2015a).

Note: The x and y percentages should be negotiated and agreed in the basin with $y > x$.

creation of an operational policy based on the zoning of reservoirs, which defines target levels as indicators of the drought stage.

These indicators function as triggers that activate the reservoir water discharge rate available for allocation during each drought stage, as well as the set of drought mitigation measures needed to reduce the impacts generated by the operational policy with safeguards.

The Jucazinho water system, located in the *agreste* region of Pernambuco, was studied, as droughts are the principal climate events in this area, and, thus, this experience can serve as a reference model for many other localities also affected by them. In this region, the reservoirs are utilized as water transport mechanisms over time and space, storing water in rainy periods, and releasing it in dry ones.

Thus, the drought indicator in this case study is based on the volume of stored water in the reservoirs. To develop it, the methodology described by Cid et al. (2014) for reservoir operation was used to construct the target levels. The historical data series from 1932 to 2008 used to simulate the different operating rules is available in the *Hydro-Environmental Plan for the Capibaribe River Basin* (Book I, Volume 1/3). The evapotranspiration data were calculated through the Penman-Monteih equation using data from the Climatological Norms from Caruaru (INMET 1992).

12.4.2 Water System Operation

Operating a reservoir means deciding how much water should be stored or released for specific ends and uses. This process can be facilitated with the utilization of information systems. This information together with mathematical models can simulate the behavior of a reservoir or system of reservoirs, associating it with a certain level of operational risk.

Campos (2009) states that a model can be understood as a set of hypotheses about the structure or behavior of a physical system, by means of which one seeks to predict or explain the properties of that system. Thus, even when not all of the characteristics of the system are known, a model is able to simulate its behavior under different conditions with a certain degree of confidence.

Reservoir simulations are essentially based on a water stock balance model. This model reproduces the hydrological performance of the reservoir on the basis of certain operational rules. Various scenarios can be utilized to compare the behavior of a reservoir, altering configurations in terms of monthly inflows and releases.

Three simulations were made for the Jucazinho reservoir, creating scenarios with different reliability values, target levels, releases, and rationing coefficients. These values were established in an *ad hoc* way, and can be modified in accordance with the decisions that are taken in the water resource planning process. However, the adoption of these values seeks to evaluate the proposed methodology, comparing it with the actual operation of the reservoir. The simulations are presented next.

1. *Target operation 1*: Release of 2177m^3 s^{-1} (value currently used in operating the reservoir), initial target levels of 0.8 * V_{max}, 0.6 * V_{max}, 0.4 * V_{max}, and 0.35 * V_{max}, and rationing coefficients of 0.9, 0.8, 0.7, and 0.5, respectively, for the target levels from 1 to 4
2. *Target operation 2*: Release of 2177 m^3 s^{-1}, guarantees of 50%, 65%, 80%, 99%, and 100%, and rationing coefficients of 0.9, 0.8, 0.7, and 0.5, respectively, for the target levels from 1 to 4
3. *Standard operation*: Operation without safeguards with an eye toward the release of 2177 m^3 s^{-1}

It should be observed that in Target Operation 1, the initial volumes in the month of June were defined for the creation of the target levels and the reservoir was only simulated, verifying its behavior. This would be a scenario should the decision makers resolve to set a minimum volume for each level of drought for the month of June. However, in Target Operation 2 the guarantees for water permanence at each storage level were established, with the objective of constructing the levels such that they approximate the proposed reliability. In this operation, an optimization algorithm was used to determine the optimum volume of water in the month of June.

In each operation, the frequency of shortfalls was analyzed, as were the severity, vulnerability, and resilience of the system and the permanence curves for each one. The frequency of *failures* was defined in this work as the number of times in which the reservoir left one stage (or level) for a lower one. *Severity* refers to the deficit in meeting a specific demand, being defined as the volume that was lacking for the reservoir to attain its target during a specific failure period. *Vulnerability* in this context measures the severity of the failures to which the system is subject, being expressed as the sum of severities.

Finally, *resilience* measures the time required by the water system to recover from a failure, should this happen. Some implications can adhere to the system if prolonged failures occur with slow recovery, as it is desirable for a system to return to a satisfactory state as quickly as possible. Hashimoto et al. (1982) define resilience as the inverse of the expected value of the average time (E) in which the system remains in failure (T_F), as presented in Equation 12.8.

$$\gamma = \frac{1}{E[T_F]} = \left[\frac{1-\alpha}{\rho}\right]^{-1} = \left[\frac{\left((1/n)\sum_{t=1}^{n}(1-Z_t)\right)}{\left((1/n)\sum_{t=1}^{n}W_t\right)}\right]^{-1} \tag{12.8}$$

where:

n is the simulation time period ($t = 1, 2, 3, \ldots, n$)

$Z_t = 0$, if a failure occurs in time t; 1, if a failure does not occur in time t

$W_t = 1$, if a failure does not occur in time t and a failure does occur in time $t + 1$; 0, the contrary case

The simulations resulted in the target levels presented in Figure 12.9. Table 12.15 presents the values of the water volumes of the reservoir in the month of July for each simulation, as well as the guarantees encountered. This is followed by a presentation of the risk analysis for each operation.

The guarantees presented represent the percentage of time in which the reservoir remained above the corresponding level. It should be emphasized that in neither simulation did the reservoir *collapse* (i.e., reach zero volume). Target Operation 2 presented better supply reliability, resulting in lower supply deficits.

The frequency of failures resulting from these operations is presented in Figures 12.10 through 12.12. It was decided only to present the failures observed for target level 1. These failures show the amount of time that the reservoir remained below its normal level. Target Operation 2 presented fewer failures at this level.

The severities of the operations are presented in Figure 12.13. Target Operation 1 presented a vulnerability of 1243 million cubic meters (hm^3), having a maximum monthly value of severity of 2.82 hm^3 and an average of 1.38 hm^3.

Target Operation 2 presented less vulnerability than Target Operation 1, with a deficit of 933 hm^3 and a maximum monthly severity of 2.83 hm^3 and an average of 1.04 hm^3/month. Finally, the standard operation assumes a lower vulnerability among the three operations, with 475.10 hm^3. Its severity

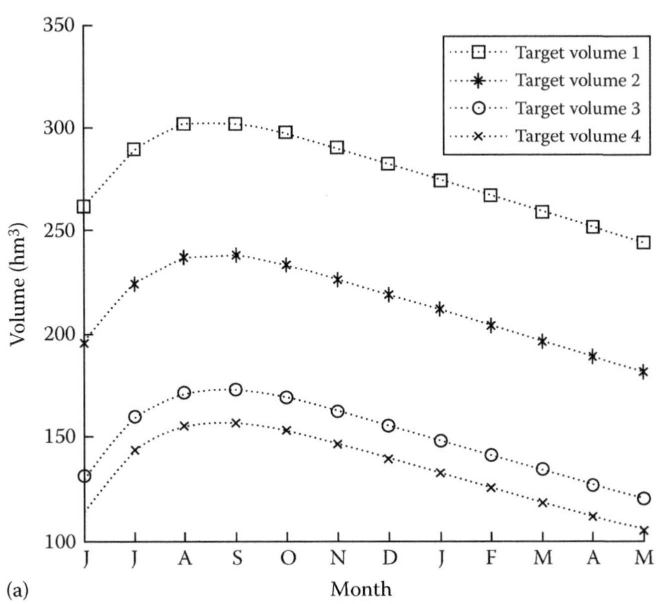

FIGURE 12.9
Target volumes established by utilization of the proposed methodology: (a) Target operation 1.
(*Continued*)

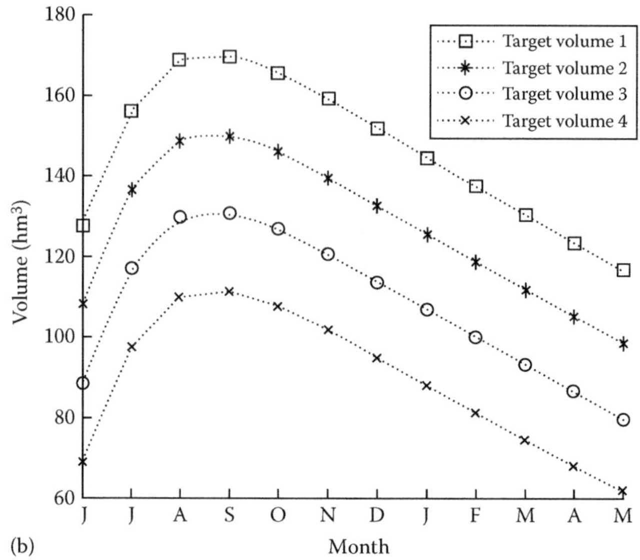

(b)

FIGURE 12.9 (Continued)
Target volumes established by utilization of the proposed methodology: (b) Target operation 2 (from June through May).

TABLE 12.15

Levels and Guarantees for the Jucazinho Reservoir for Target Operations 1 and 2

Target Level	Target Operation 1: Initial Volume Month of June (hm³)	Target Operation 1: Guarantee (%)	Target Operation 2: Initial Volume Month of June (hm³)	Target Operation 2: Guarantee (%)
1	261.62	11.55	127.56	44.33
2	196.22	31.78	107.93	54.33
3	130.81	64.88	88.29	65.78
4	114.46	73.44	68.66	75.66

value is equal to 5.64 hm^3 (2177 $m^3\ s^{-1}$), corresponding to not meeting demand during the period in which the reservoir enters collapse. The average for this operation presented a value of 0.53 hm^3/month.

The resilience time of the operations is presented in Table 12.16. Again, the resilience measures the time the water system needs to recover from a failure, should this occur.

The standard operation presented an average resilience time of 7.07 months in order to recover from a state of collapse, while under the other operations the reservoir did not attain this state. Target Operation 2 presented resilience times lower than Target Operation 1 for all levels except number 4.

The monthly accumulations and discharges from the reservoir were also analyzed. Figures 12.14 and 12.15 present the permanence curves for the two

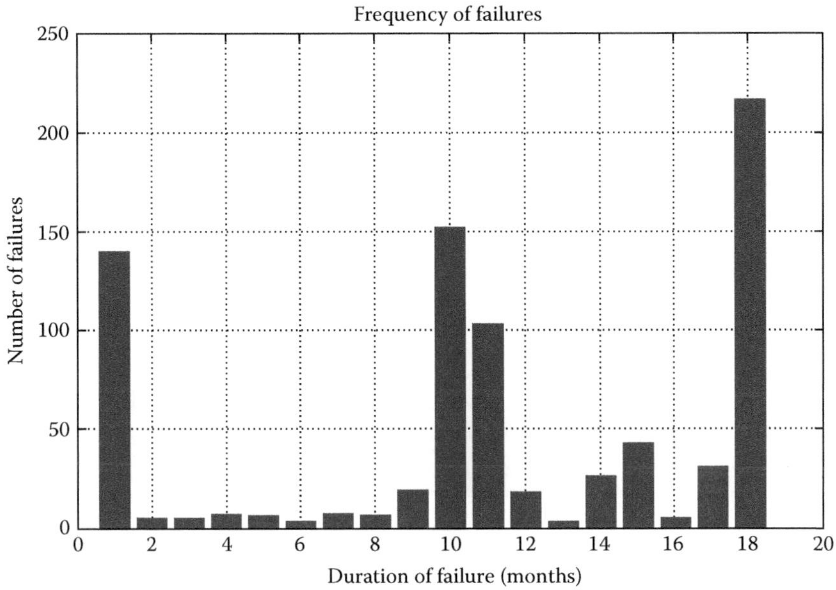

FIGURE 12.10
Target operation 1: Histogram of the quantity of failures at target level 1. At this level, failures occurred during 796 months, with the maximum duration of 217 months and minimum of 3 months.

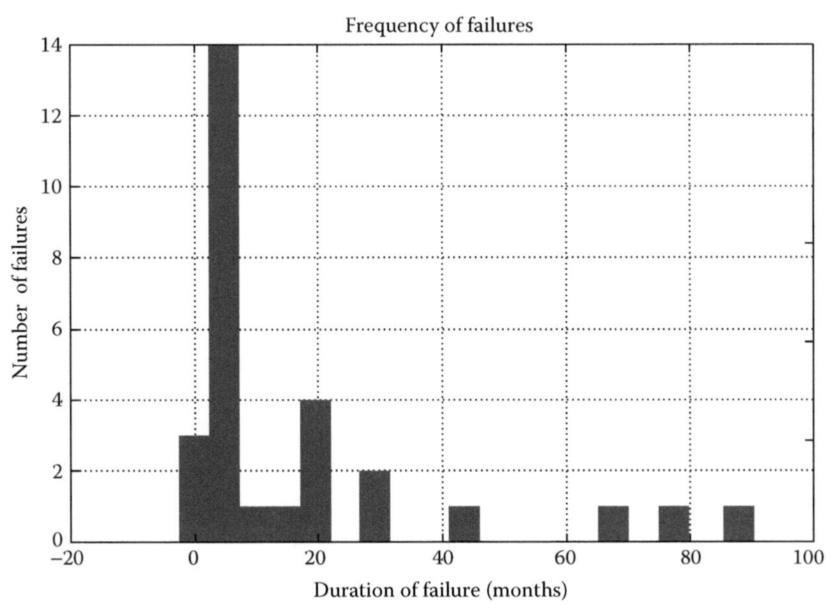

FIGURE 12.11
Target operation 2: Histogram of the quantity of failures at target level 2. At this level, failures occurred during 501 months, with the maximum duration of 91 months and minimum of 2 months.

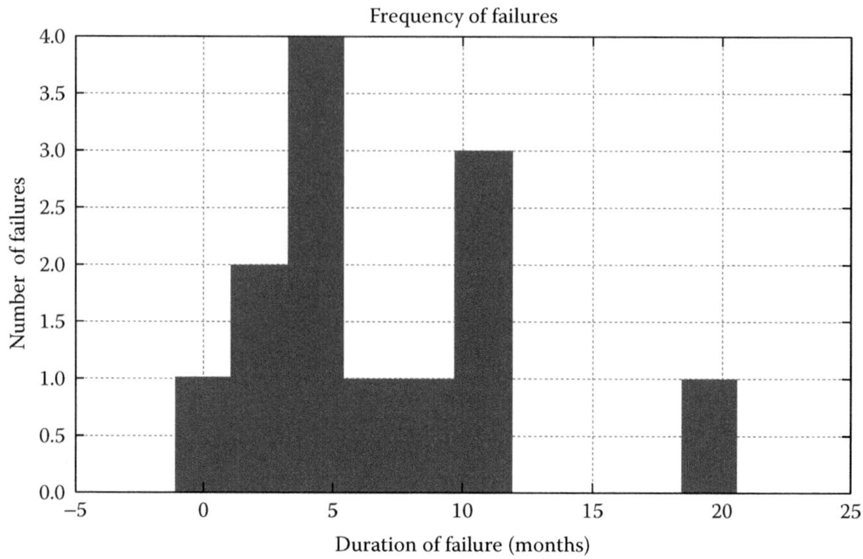

FIGURE 12.12
Standard operation: This operation presented 92 failures of collapse of the reservoir, having one that presented a maximum duration of 20 months and one that presented a minimum duration of 1 month.

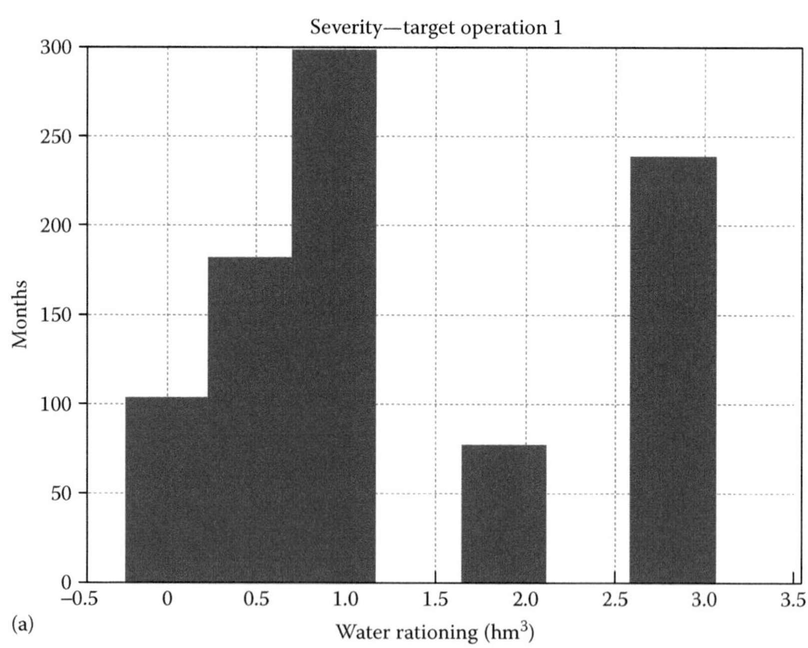

FIGURE 12.13
Histogram of severity of the operations: (a) Target 1. *(Continued)*

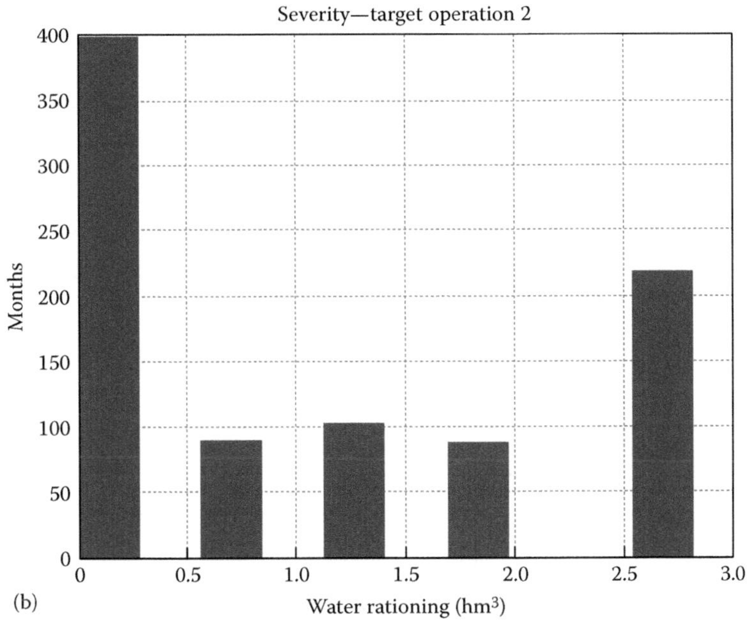

FIGURE 12.13 (Continued)
Histogram of severity of the operations: (b) Target 2.

TABLE 12.16
Resilience Time of Target Operations 1 and 2 and Standard Operation

	Resilience Time (Months)		
Target Level	**Target Operation 1**	**Target Operation 2**	**Standard Operation**
1	44.2	17.2	–
2	24.56	14.2	–
3	9.3	8.8	–
4	7.96	8.42	–
Dead storage	0	0	7.07

variables considered. The Standard Operation presented the lowest accumulation volume among the three operations, reaching a zero level approximately 18% of the time.

With Target Operation 1, it is possible to accumulate more water in the reservoir because the rule allows for the water to stay there because of longer rationing period. Target Operation 1 releases volumes equal to 5.64 hm^3 during approximately 90% of the period, but during 10% of the period it is unable to meet demand. Target Operations 1 and 2 guarantee a minimum supply of 2.15 hm^3 25% of the time. However, Target Operation 2

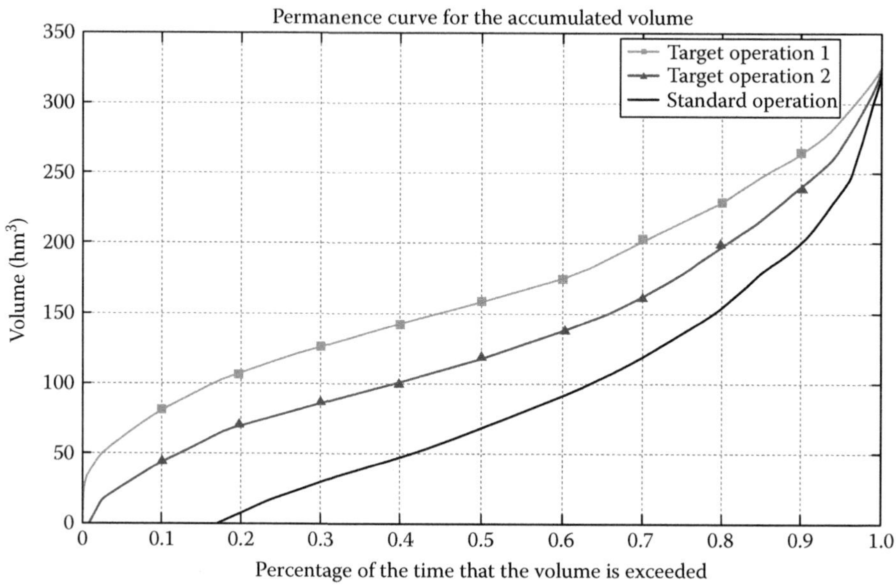

FIGURE 12.14
Permanence curve for the accumulated volume for the three operations.

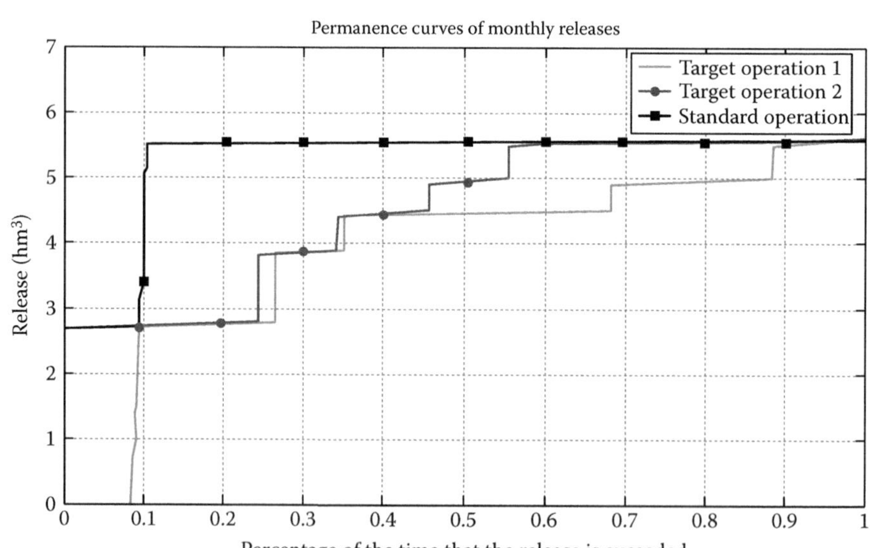

FIGURE 12.15
Permanence curves of monthly releases for the three operations.

results in larger releases during the majority of the time compared with Target Operation 1.

12.4.3 Actions

Target Operation 1 was selected as the scenario for operation of the Jucazinho reservoir for this drought preparedness plan. Table 12.17 presents the target levels for the four most severe drought categories: moderate drought, severe drought, extreme drought, and exceptional drought. Table 12.18 indicates the response targets defined by the planners for each of these stages.

Each one of these stages is associated with a set of drought mitigation measures. These measures were classified in three operational, organizational, and monitoring groups. The first has the purpose of reducing water demand by means of operational actions such as the reduction in nocturnal pressure in urban networks.

As organizational measures, the following can be cited: (1) the preparation and approval of decrees and resolutions necessary for effectiveness of the actions; (2) evaluating the quantity of resources necessary for the utilization of emergency water supply plans and preparation of the documentation necessary for resource requests; and (3) preparation of food supply plans for the localities affected by the drought.

TABLE 12.17

Reservoir Target Level For Each Drought Stage

Moderate drought	261.6	284.9	296.4	299.0	297.0	289.8	282.0	274.1	266.2	258.5	250.8	243.5
Severe drought	196.2	219.7	231.5	234.3	232.5	225.7	218.2	210.6	203.0	195.7	188.3	181.2
Extreme drought	130.8	154.6	166.6	169.7	168.2	161.7	154.6	147.4	140.2	133.2	126.2	119.4
Exceptional drought	114.5	138.3	150.3	153.5	152.1	145.8	138.7	131.6	124.5	117.6	110.7	104.0

Source: Research data.

TABLE 12.18

Drought Stages, Triggers, and Response Targets Defined by the Working Group

Drought Scenarios		
Stage	**Drought Triggers**	**Response Targets**
Moderate drought	Below alert target level	10% reduction in consumption
Severe drought	Below drought target level	20% reduction in consumption
Extreme drought	Below severe drought target level	30% reduction in consumption
Exceptional drought	Below extreme drought target level	50% reduction in consumption

Monitoring measures include (1) monitoring the volume of water in the reservoirs; (2) monitoring various hydro–meteorological variables; and (3) monitoring water demand.

These sets of actions were incorporated into a decision support system (DSS). In this system, with the volume of water in the reservoir, the planner identifies the potential amount of water release for allocation, the drought stage, and the measures associated with it.

12.5 Urban Drought Preparedness Plan

12.5.1 Context

As with the river basin and the hydrosystem presented in Sections 12.3 and 12.4, respectively, the elaboration of an urban drought preparedness plan requires the development of a monitoring and early warning system, the evaluation of impacts and vulnerability, and the definition of preparedness, mitigation, and response actions and/or programs. As indicated earlier, the monitoring is associated with the use of appropriate indices/indicators and triggers, together with the development of a DSS.

In this context, the metropolitan region of Fortaleza (FMR) was the focus of the application of a specific vulnerability assessment methodology (pressure-state-impact-response, or PSIR). This region is composed of numerous municipalities, including: Aquiráz, Cascavél, Caucaia, Chorozinho, Eusébio, Fortaleza, Guaiuba, Horizonte, Itaitinga, Maracanaú, Pacatuba, Pacajús, Paracurú, Paraipaba, Pindoretama, São Gonçalo do Amarante, São Luis do Curú, and Trairí. Together, they have a population of more than 3.7 million, with more than 60% of the total residing in the municipality of Fortaleza, which is also the state capital of Ceará. This population concentration is not static, and has increased over time as the result of factors like urbanization and industrialization, elevating the demand for water, which is now a scarce commodity.

According to data from the Ceará Water Resource Management Company (COGERH), in 2015, 87.92% of the demand for water in the FMR is being met by the Castanhão reservoir. This fact evidences the significant dependence by one part of the state on water drained from another to meet its needs for urban use. But the water crisis situation is reflected in the low storage volumes contained in the reservoirs throughout the state, which in mid-2015 were at only 15.6%* of their normal capacity. These data reveal the need for the elaboration of an urban drought preparedness plan capable of

* Data drawn from the Hydrological Portal of the Government of Ceará.

anticipating the critical situation, based on a monitoring and early warning system conceived under the lens of a proactive logic. The steps for the elaboration of a plan with this purpose applied to the Fortaleza metropolitan region are presented next.

12.5.2 Monitoring and Early Warning

Drought monitoring and early warning can be undertaken by constructing indicators that reveal the status of a drought in an urban environment. These indicators can be built according to the methods presented earlier in this chapter and the decision maker can choose to use only one or several of such indicators in the drought preparedness plan.

Signaling the severity status of a drought occurs by means of the ranges of the values of the specific indicator or indicators utilized. Thus, each time the upper limit of the range—or threshold—is passed, a higher drought status is defined and an appropriate set of actions is triggered. The set of actions associated with a particular drought severity status has as its objective to achieve a water supply increase target, reduce water demand, adapt the population, and mitigate impacts and conflicts resulting from the drought.

In addition to monitoring and early warning, another significant step in drought preparedness planning is the evaluation of existing vulnerabilities in the urban environment. This evaluation is a tool capable of identifying elements that trigger the development of coordinated short-, medium-, and long-term measures within a proactive logic.

12.5.3 Vulnerability Analysis

Gallopín (2003) defends the thesis that vulnerability is a function of the response capacity and sensitivity of a system that faces disturbances, and should not be conceptually associated only with negative ones. For him, there is a positive dimension when the transformation experienced is beneficial. But he recognizes that the majority of such disturbances result in vulnerabilities due to negative impacts on the systems and populations in question, although these may occur with different intensities.

Vulnerability is a concept that is associated with the term *resilience*, which is fundamental in order to overcome impasses, problems, and impacts caused by vulnerability. However, the perception and understanding of vulnerability should be based on analyses that make it possible to identify what is causing it and its consequences as part of a strategy that is capable of formulating responses that are appropriate for the situation assessed. This is derived from the importance of methodologies that guarantee that vulnerability analyses are undertaken.

It is in this context that one can apply the PSIR methodology, which permits identification of the factors that generate pressure on a particular sector, modifying its state, and resulting in impacts that demand specific responses. This methodology was conceived by the Organization of Economic Cooperation and Development (OECD) and was complemented by the United Nations Environment Program (UNEP) in 2007.

With the adoption of the PSIR methodology,* one seeks to identify the mechanisms and strategies capable of strengthening the adaptation capacity of the water resource system in the face of climate changes or variations that occur at the local level caused by climate extremes, favoring the linkage between the actual or potential impact and elaboration of adaptation policies, as indicated by Pahl-Wostl (2007). In this way, pressure is considered in the form of climate factors and polluting actions that exert negative pressure on the water resource sector, depleting it and/or leading to its degradation. These pressures create the critical situations that generate impacts on the water system in question, requiring responses, which take the form of strategies conceived on the basis of a proactive logic in the context of adaptive management.

For purposes of this study, this methodology was adapted to assess the vulnerability of the water treatment plant (WTP) for the Fortaleza metropolitan region, whose application can be visualized in Figure 12.16. This plant is located at the Gavião reservoir and presents a treatment capacity of 1.5 $m^3 s^{-1}$. It was initially projected to use a conventional treatment technology, but this was substituted in 1995 by direct descendant filtration. This change increased the WTP's treatment capacity by 25%.

The diagram presented in Figure 12.16 synthesizes the dimensions of the vulnerability analysis method applied to the WTP. It should be observed that the reduction in the inflow and elevated evaporation characteristic of the semiarid region in which the plant is situated, and the discharge of pollutants, continually increasing over time as a result of the installation of new industrial enterprises, are considered as pressure indicators, as they can result in significant changes in the treatment system generating both economic and social harm.

These pressures imposed on the WTP can lead to a state of reduced water storage levels that, in turn, result in a reduction of the water released for consumption. There is a higher concentration of conservative elements (chlorine, phosphorous), which increases the salinity of the water and the emergence of filamentous algae. In addition, there is an increase in the biochemical oxygen demand (BOD) due to the higher quantity of organic matter in the water, whereby extreme levels can provoke eutrophication of reservoirs, having a direct connection with an increase in cyanobacteria and water turbidity, resulting in the depletion of dissolved oxygen and, consequently, an increase in anaerobic processes, odor, and toxins.

* This is one of the variations of the pressure-state-response (PSR) methodology. Other methodologies that include the energizing force such as FSR (force-state-response) and FPSIR (force-pressure-state-impact-response), could also be utilized.

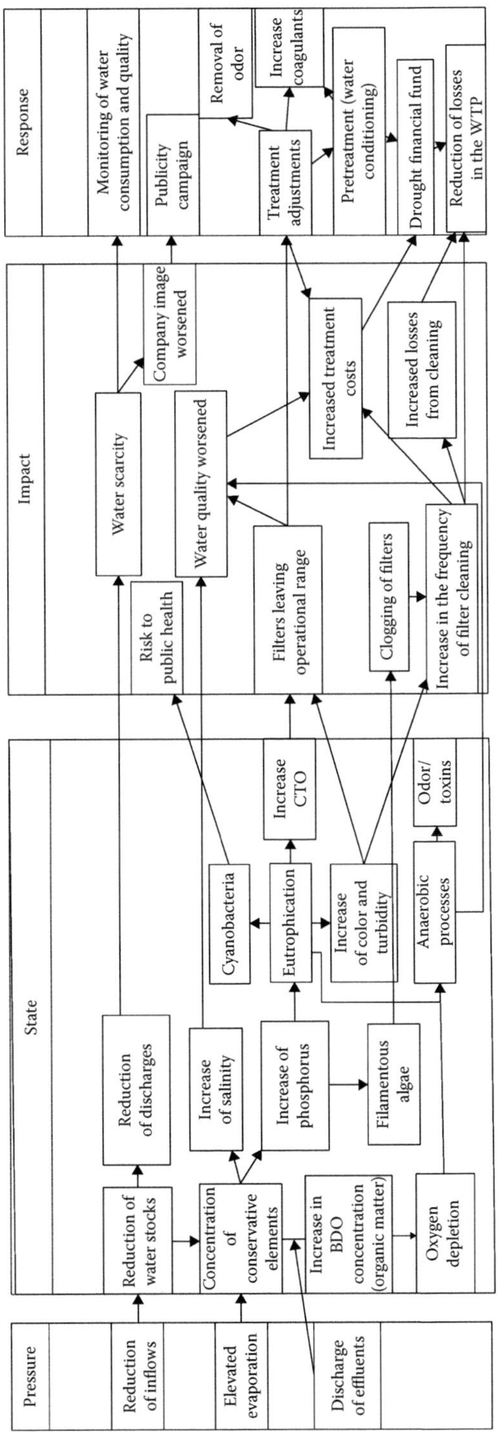

FIGURE 12.16
Application of the PSIR methodology to the Fortaleza water treatment plant.

In the face of this situation, the WTP supplies a lower volume of water, reducing the life of the filters due to a decrease in their operation between cleanings and consequent clogging and generating a public health risk. This requires greater frequency of cleaning and, thus, increased treatment costs, and risks greater contamination due to the presence of cyanobacteria and toxins in the water.

The response dimension corresponds to the actions taken by the water supply company to attenuate the effects of the drought. In synthesis, this consists of (1) increased monitoring of water consumption and quality in order to interrelate water availability with the socio economic and geo environmental peculiarities of the region under study; (2) adjustments in the treatment process and water purification, which are the main ways of reducing water contamination; and (3) creation of a fund, because when droughts occur, there is an increase in the expenses for water management, monitoring, and treatment as well as to reduce apparent and actual losses related to the nonauthorized consumption of water (i.e., fraud and cadastral omissions), imprecision of the water meters, or leaks in the aqueducts and distribution networks, respectively.

12.5.4 Action Plan

In the urban drought preparedness management plan it is necessary to develop a set of actions to be triggered in order to mitigate the effects of the drought and promote adaptation to it. These actions involve environmental, social, and institutional measures as well as improvements to the system, monitoring, and ensuring that the objectives of the management model are met. They should be the result of discussions among the institutions directly or indirectly responsible for the urban water supply system. The set of social actions, for example, should seek to reduce water demand and to provide adequate and useful information to consumers. Those of an environmental character, in turn, have the purpose of reducing significant impacts resulting from the use of water resources.

The institutional measures seek to integrate the entities with responsibility for addressing the drought. Several questions should be observed with respect to the coordination of the technical team in relation to the distribution of tasks and the boundaries between them. This is possible through a horizontal communication process, in which the various stakeholders involved in drought preparedness planning have access and the ability to process relevant information so that they will be capable of pulling the triggers in an efficient manner, thereby contributing to the reduction of impacts that could affect cities and their inhabitants.

In implementing these sets of actions, it is important that three dimensions be observed: (1) the normative framework; (2) management aspects; and (3) establishment of an agenda with the regulatory agency.

In relation to the normative framework, implementation of an action plan requires formation of a specific team to execute and monitor it. The normative framework should facilitate the implementation of operational rules throughout the urban water supply system in order to support management of the drought preparedness plan.

As concerns the management aspect, there is a need to constitute a drought management group, as well as to define the respective attributions and responsibilities of each of its members. At each drought stage, the team will plan, in a more detailed form, the action to be taken in the next one, as part of a thoughtful planning process within a proactive approach. In this dimension also, a decision making agenda is defined, including the frequency of discussions and meetings to be undertaken by the management group.

The action plan was conceived through the definition of an integrated set of action *classes*, in which a typology, description, indication of responsible institution(s), and priority level was determined for each one. Thus, for example, nocturnal interruptions are a type of action that are inserted in the *system management and operations* class, having as their objective to reduce water consumption and avoid the utilization of new water supply sources that can often result in large-scale environmental harm. This drought stage measure is in priority 1, which refers to necessary actions.

It is important to emphasize that for each stage there are a set of actions that can be maintained, added to, or simply are not necessary for that specific stage. These actions were classified as necessary (priority 1), important (priority 2), and complementary (priority 3). Clearly, the priority level also changes as a function of changes in the drought situation. An example of this plan can be observed in Table 12.19 that refers to the *drought* stage.

12.5.5 Decision Support System

To assist decision makers with respect to drought early warning and monitoring within the urban drought preparedness plan, the development of a DSS is proposed. The DSS has the purpose of supporting urban water supply management, presenting the planner with a view of the state of the reservoir and alternatives with respect to the actions to be taken in a drought situation, and should be understood as part of a broader drought management planning process.

The DSS can be divided into two stages: (1) monitoring and early warning and (2) drought response. The first stage is based on inserting the volume of the reservoir in the system and observing the drought stage identified by it. The drought stage detected is associated with a value of discharge defined on the basis of the simulation of reservoir operation and mass balance and a set of actions to be taken by the water resources managers so that the drought does not attain a higher state of severity.

With this, the DSS permits planners and decision makers to accompany on a daily basis the state of the reservoir for urban water supply and take decisions to mitigate the drought.

TABLE 12.19

Class of Actions with Indication of Typology, Description, Responsible Institution, and Priority Level for the Drought Stage

Stage	Class	Action Type	Description	Responsible Institution	Priority Level
Drought	Monitoring and preventive measures	Periodic evaluation of consumption by locality; quantitative and qualitative availability	Assess the consumption problem in each locality, interrelating water availability with their socio economic and geo environmental characteristics (tariff increase/reduction, rationing).	CAGECE/ COGERH	1
		Monitoring the quality of raw water	Monitoring is an intense activity to accompany, oversee, and simultaneously evaluate the environment in order to diagnose the situation.	COGERH	1
		Monitoring the quality of treated water	This action seeks to constantly monitor water quality and its use for human consumption, avoiding the dissemination of water-borne diseases.	CAGECE	1
		Monitoring hydro meteorological parameters	This action is necessary to assess reservoir recharge.	COGERH/ FUNCEME	1
		Inventory of alternative water supply sources	The inventory of these sources is significant for the water supply to the population, as well as to maximize the efficiency of the water system.	SRH/COGERH	1

12.6 Final Observations

Increases in the number of drought events, their severity, and their impacts have led a growing number of governments to adopt a more proactive approach to drought management, in an effort to reduce their short-term impacts and longer term vulnerability. The development of drought policies that promote risk management and elaboration of contingency plans are examples of a change of paradigm by governments in their approaches to drought management.

Drought preparedness requires definition of the various stages of severity of a drought and of an appropriate indicator or indicators. This also requires greater coordination within and among levels of government, assessment of impacts, and definition of mitigation and response strategies and actions.

In the elaboration of a drought preparedness plan, the specificities of the hydro system for which planning is to occur should be observed. Development of a monitoring system, with a clear definition of drought indicators and thresholds between drought stages is a fundamental step for the development of a drought preparedness plan. Identification of the vulnerabilities of the water system in question to the drought contributes to the definition of effective measures to mitigate drought impacts. Associated with this, the planning process should incorporate all relevant stakeholders for the conception and implementation of the plans.

Bibliography

Agência Nacional de Águas. 2014. *PRH Piranhas-Açu, Plano da Bacia Hidrográfica do Rio Piranhas-Açu*. Brasil.

Agência Nacional de Águas. 2015. *Boletim diário de acompanhamento do Sistema Cantareira de 23/set/2015*.

Amanajás, Jonathan C. 2008. *Uso do método de contingência para análise da eficácia de previsão da precipitação pluviométrica do modelo ETA para o município de Macapá-AP em 2007*. Federal University Amapá, Macapá, Brazil.

Andersen, Lykke E., Clemens Breisinger, Daniel Mason-D'Croz, Luis Carlos Jemio, Claudia Ringler, Richard Robertson, Dorte Verner, and Manfred Wiebelt. 2014. An assessment of climate change impacts and household resilience for Brazil, Mexico, and Peru. IFPRI Discussion Paper 01390, International Food Policy Research Institute, Washington, DC.

Andersen, Lykke E., Clemens Breisinger, Daniel Mason-D'Croz, Luis Carlos Jemio, Claudia Ringler, Richard Robertson, Dorte Verner, and Manfred Wiebelt. 2015. Agriculture, incomes, and gender in Latin America by 2050: An assessment of climate change impacts and household resilience for Brazil, Mexico, and Peru. IFPRI Food Policy Report Series, International Food Policy Research Institute, Washington, DC.

Andersen, Lykke E. and Dorte Verner. 2010. *Simulating the Effects of Climate Change on Poverty and Inequality*. World Bank, Washington, DC.

Andreu, Joaquin and A.S. Solera. 2005. Methodology for the analysis of drought mitigation measures in water resource systems. In: Andreu, Joaquin, Giuseppe Rossi, Federico Vagliasindi, and Alicia V. Mayorga (eds.), *Drought Management and Planning for Water Resources*. Taylor & Francis Group, Boca Raton, FL, pp. 133–168.

Antwi-Agyei, Phillip, Evan D.G. Fraser, Andrew J. Dougill, Lindsay C. Stringer, and Elisabeth Simelton. 2012. Mapping the vulnerability of crop production to drought in Ghana using rainfall, yield and socioeconomic data. *Applied Geography*, 32(2): 324–334.

Assad, Eduardo D., Susian C. Martins, Napoleão E. de Macêdo Beltrão, and Hilton S. Pinto. 2013. Impacts of climate change on the agricultural zoning of climate risk for cotton cultivation in Brazil. *Pesquisa Agropecuária Brasileira*, 48(1). Available from: http://www.scielo.br/scielo.php?script=sci_arttext&pid=S0100-204X2013000100001&lng=en&tlng=en.%2010.1590/S0100-204X2013000100001.

Balzam, Guilherme. 2010. Rios Solimões e Amazonas têm maior seca da história. *UOL Noticias*, October 22. Available from: http://noticias.uol.com.br/cotidiano/ultimas-noticias/2010/10/22/rios-solimoes-e-amazonas-tem-maior-seca-da-historia.htm.

Banco do Nordeste do Brasil. 2005. Fundação Cearense de Meteorologia e Recursos Hídricos. *Redimensionamento da Região Semiárida do Brasil*, BNB/FUNCEME, Fortaleza, Brazil.

Bastos, Paulo, Matias Busso, and Sebastian Miller. 2013. Adapting to climate change: Long-term effects of drought on local labor markets. IDB Working Paper No. IDB-WP-466, World Bank, Washington, DC.

Bazza, Mohamed. 2001. Inference of a drought mitigation action plan. Expert consultation and workshop on drought preparedness and mitigation in the near east and the Mediterranean. Aleppo, Syria, May 27–31, 2001, FAO, Regional Office for the Near East, Cairo, Egypt. Available from: ftp://193.43.36.93/docrep/fao/010/ai149e/ai149e.pdf.

Bhattacharya, Sumana and Aditi Das. 2007. Vulnerability to drought, cyclones and floods in India. The BASIC Project, Paper 9, Winrock International, India.

Bordieu, Pierre. 1989. *Razões práticas: Sobre a teoria da ação.* Papirus, Campinas, Brazil.

Braga, Renato. 2004. *História da Comissão Científica de Exploração.* Edições Demócrito Rocha, Fortaleza, Brazil.

Brant, Simone. 2007. Assessing vulnerability to drought in Ceará, Northeast Brazil. Master's Thesis, University of Michigan, Ann Arbor, MI.

Brasil. 1934a. *Constituição da República dos Estados Unidos do Brasil (de 16 de julho de 1934).*

Brasil. 1934b. *Decreto N° 24.643, de 10 DE Julho de 1934, o Código de Águas.*

Brasil. 1985. Instituto Brasileiro de Geografia e Estatística. *Atlas Nacional do Brasil, Região Nordeste,* IBGE, Rio de Janeiro, Brazil.

Brasil. 1988. *Constituição da República Federativa do Brasil.*

Brasil. 2006. Companhia de Pesquisas de Recursos Minerais. *Mapa Geodiversidade do Brasil,* escala 1:2.500.000, Legenda Expandida, ilustrada, CD-ROM/SIG, Serviço Geológico do Brasil - CPRM, Secretaria de Geologia, 12 Mineração e Transformação Mineral - SGM, Ministério de Minas e Energia - MME, Brasília, Brazil.

Brasil. 2006. Instituto Brasileiro de Geografia e Estatística. *Mapa de Unidades de Relevo do Brasil,* escala 1:5.000.000, IBGE, Rio de Janeiro, Brazil.

Brasil. 2007. Ministério do Meio Ambiente - MMA. *Atlas das áreas suscetíveis à desertificação do Brasil.* Ministério do Meio Ambiente, Brasília, Brazil.

Brasil. 2007. Instituto Brasileiro de Geografia e Estatística - IBGE. *Malha Municipal Digital do Brasil: situação em. Rio* de Janeiro, Brazil. Available from: ftp://geoftp.ibge.gov.br/malhas_digitais/. Last Accesed on November, 2015.

Burgess, Robin, Olivier Dêschenes, Dave Donaldson, and Michael Greenstone. 2011. Weather and death in India: Mechanisms and implications for climate change. Manuscript, Department of Economics, Massachusetts Institute of Technology, Cambridge, MA.

Burrough, Peter A. and Rachel A. McDonnell. 1986. *Principles of Geographical Information Systems.* Oxford University Press, New York.

Calvosa, Chiara, Delgermaa Chuluunbaatar, and Katiuscia Fara. 2009. Livestock and climate change. Livestock Thematic Papers. Tools for project design. The International Fund for Agricultural Development (IFAD), pp. 1–20. Available from: http://www.ifad.org/lrkm/factsheet/cc.pdf.

Campello, Tereza. and Marcelo Côrtes Neri. 2013. *Programa Bolsa Família: uma década de inclusão e cidadania.*

Campos, José Nilson B. 2009. *Lições em Modelo e Simulação Hidrológica.* ASTEF, Fortaleza, Brazil.

Campos, José Nilson B. 2014. Secas e políticas públicas no semiárido: Ideias, pensadores e períodos. *Estudos Avançados,* 28(82): 65–88. Instituto de Estudos Avançados, Universidade de São Paulo, São Paulo, Brazil.

Campos, José Nilson B. 2015. Paradigms and public policies on drought in northeast Brazil: A historical perspective. *Environmental Management,* 55(5): 1052–1063.

Carvalho, Otamar. 2013. *Estudos Referentes ao Diagnóstico da Política Nacional de Secas no Brasil.* Ministério da Integração Nacional, Brasília, Brazil.

Centro de Gestão e Estudos Estratégicos (CGEE). 2016. *Desertificação, Degradação da Terra e Seca no Semiárido Brasileiro.* CGEE, Brasília, Brazil.

Chimeli, Ariaster B., Francisco de Assis de Souza Filho, Marcos C. Holanda, and Francis C. Petterini. 2008. Forecasting the impacts of climate variability: Lessons from the rainfed corn market in Ceará, Brazil. *Environment and Development Economics* 13(2): 201–227.

Cid, Daniel A.C., Francisco de A. de Souza Filho, Luiz Martins de Araújo Junior, and Samiria M. O. da Silva. 2014. Uso de simulação para definição de níveis metas de operação para o reservatório Jucazinho/PE. In *XII Simpósio de Recursos Hídricos do Nordeste,* Natal, Rio Grande do Norte, Brazil.

Colorado Water Conservation Board. 2010. Municipal drought management plan guidance document. AMEC Earth & Environmental, Boulder, CO.

Colorado Water Conservation Board. 2013. The Colorado drought mitigation and response plan August. Department of Natural Resources and AMEC Environment and Infrastructure, Colorado Water Conservation Board, Boulder, CO.

Confederación Hidrográfica del Ebro (CHE). 2007. *Plan especial de actuación en situaciones de alerta y eventual sequia en la cuenca hidrográfica del ebro.* CHEMARM. Zaragoza.

Confederación Hidrográfica del Júcar. 2007. *Plan especial de alerta y eventual sequia en la Confederacion Hidrografica del Jucar.* Ministerio de Medioambiente. Valencia.

Cooper-Johnston, Ross. 2000. *El Niño. The Weather Phenomenon That Changed the World.* Hodder & Stoughton, Coronet Books, London, UK.

Cunha, Rita L.A. 2008. Definição de cenários de referência para a avaliação dos impactos das secas. Masters' Thesis, Mestrado Integrado em Engenharia Civil—2007/2008—Departamento de Engenharia Civil, Faculdade de Engenharia da Universidade do Porto, Porto, Portugal.

Davis, Mike. 2000. *Late Victorian Holocausts: El Niño Famines and the Making of the Third World.* Verso Books, London, UK.

Denver Water. 2014. *Drought Response Plan.* Denver, CO. 22p.

Deschênes, Olivier, and Michale Greenstone. 2007. The economic impacts of climate change: Evidence from agricultural output and random fluctuations in weather. *American Economic Review,* 97(1): 354–385.

Ding, Ya, Michael Hayes, and Melissa Widhalm. 2011. Measuring economic impacts of drought: A review and discussion. *Disaster Prevention and Management,* 20(4): 434–446.

Dinkelman, Taryn. 2013. Can spatial mobility insure families against long-term impacts of economic shocks? Evidence from drought and disability in South Africa. Dartmouth College, Hanover, NH.

Engle, Nathan L. 2012. The role of drought preparedness in building and mobilizing adaptive capacity in states and their community water systems. *Climatic Change.* DOI: 10.1007/s10584-012-0657-4.

Estrela, Teodoro, Aránzazu Fidalgo, and Miguel A. Pérez. 2006. Droughts and the European water framework directive: Implications on Spanish river basin districts. In: Andreu, Joaquin, Giuseppe Rossi, Federico Vagliasindi, and Alicia V. Mayorga (eds.), *Drought Management and Planning for Water Resources.* CRC Press, Taylor & Francis Group, Boca Raton, FL, Chapter 6, pp. 169–191.

Felsenthal, Dan S. 1980. Applying the redundancy concept to administrative organizations. *Public Administration Review*, 40: 247–248.

Fernandes, Erick C.M., Ayat Soliman, Roberto Confalonieri, Marcello Donatelli, and Francesco Tubiello. 2012. *Climate Change and Agriculture in Latin America, 2020–2050: Projected Impacts and Response to Adaptation Strategies*. World Bank, Washington, DC.

Food and Agriculture Organization and National Drought Mitigation Center. 2008. *The Near East Drought Planning Manual: Guidelines for Drought Mitigation and Preparedness Planning*. University of Nebraska, Lincoln, NE. Available from: http://drought.unl.edu/portals/0/docs/NearEastDroughtPlanningManual.pdf.

Fuchs, Brian, Mark Svoboda, Jeff Nothwehr, Chris Poulsen, William Sorensen, and Nathaniel Guttman. 2012. A New National Drought Risk Atlas for the U.S. from the National Drought Mitigation Center. Available from: http://www.clivar.org/sites/default/files/Fuchs.pdf.

Gallopín, Gilberto C. 2003. A systemic synthesis of the relations between vulnerability, hazard, exposure and impact, aimed at policy identification. In: Economic Commission for Latin American and the Caribbean (ECLAC) (ed.), *Handbook for Estimating the Socio-Economic and Environmental Effects of Disasters*. LC/MEX/G.5, ECLAC, Washington, DC.

Gallopín, Gilberto C. 2006. Linkages between vulnerability, resilience, and adaptive capacity. *Global Environmental Change*, 16: 293–303.

Giddens, Anthony. 2000. *O mundo na era da globalização*. Presença, Lisbon, Portugal.

Godoy, Arilda S. Mar./Abr. 1995. Introdução à pesquisa qualitativa e suas possibilidades. Uma revisão histórica dos principais autores e obras que refletem esta metodologia de pesquisa em Ciências Sociais. *Revista de Administração de Empresas São Paulo*, 35(2): 20–29.

González, A. 2012. Regional strategic framework for climate risk management in the agricultural sector of the Central American Dry Corridor. Food and Agriculture Organization of the United Nations (FAO). Available from: http://coin.fao.org/coin-static/cms/media/14/13590441298720/marco_estratgico_corredor_seco-.pdf.

González, Francisco C. and José De Castro Morcillo. 2007. *Guía para la elaboración de planes de emergência por sequía em sistemas de abastecimiento urbano*. Ministerio de Medio Ambiente, Associación Española de abastecimentos de agua y saneamento, 83p.

Gray, Clark and Valerie Mueller. 2012. Drought and population mobility in rural Ethiopia. *World Development*, Elsevier, 40(1): 134–145.

Gutiérrez, Ana Paula A., Nathan L. Engle, Erwin De Nys, Carmen Molejón, and Eduardo S. Martins. June 2014. Drought preparedness in Brazil. *Weather and Climate Extremes*, 3: 95–106.

Hall, Anthony L. 1978. *Drought and Irrigation in North-East Brazil*. Cambridge University Press, Cambridge.

Hashimoto, Tsuyoshi, Jery R. Stedinger, Daniel P. Loucks. 1982. Reliability, resilience and vulnerability criteria for water resource system performance evaluation. *Water Resources Research*, 18(1): 14–26.

Heim Jr., Richard R. 2002. A review of twentieth-century drought indices used in the United States. *Bulletin of the American Meteorological Society*, 83: 1149–1165.

Hirschman, Albert O. 1963. *Journeys toward Progress: Studies of Economic Policy Making from Latin America*. Twentieth Century Fund, New York.

Instituto Brasileiro de Geografia e Estatistica (IBGE). 2004. *IBGE Releases Previously Unseen Profile of Food Security in Brazil*. Press Release, May 17, 2006. Available from: http://www.ibge.gov.br/english/presidencia/noticias/noticia_visualiza.php?id_noticia=600&id_pagina=1.
Instituto Nacional de Meteorologia (INMET). 1992. *Normais Climatológicas (1961–1990)*. Brasília, Brazil.
Intergovernmental Panel on Climate Change (IPCC). 2013. *Climate Change 2013: The Physical Science Basis: Working Group I Contribution to the 5th Assessment Report of the Intergovernmental Panel on Climate Change*, IPCC, Geneva, Switzerland. Cambridge University Press, Cambridge, UK.
Intergovernmental Panel on Climate Change (IPCC). 2014. *Climate Change 2014. Impacts, Adaptation, and Vulnerability. Working Group II contribution to the 5th Assessment of the Intergovernmental Panel on Climate Change*, IPCC, Geneva, Switzerland. Cambridge University Press, Cambridge UK.
Jayachandran, Seema. 2006. Selling labor low: Wage responses to productivity shocks in developing countries. *Journal of Political Economy*, 114(3): 538–575.
Keyantash, John and John A. Dracup. 2002. The quantification of drought: An evaluation of drought indices. *Bulletin of the American Meteorological Society*, 83(8): 1167–1180. DOI: 10.1175/1520–0477.
Khan, Ahmad S., José A.N. da Cruz, Lucia M.R. Silva, and Patricia V. P. S. Lima. 2005. Efeito da seca sobre a produção, a renda e o emprego agrícola na microregião geográfica de Brejo Santo e no Estado do Ceara. *Revista Economica do Nordeste*, Fortaleza, 36(2): 242–262.
Lackstrom, Kirsten, Amanda Brennan, Daniel Ferguson, Mike Crimmins, Lisa Darby, Kirstin Dow, Keith Ingram et al. 2013. The missing piece: Drought impacts monitoring. National Drought Mitigation Center. Available from: http://www.drought.gov/media/pgfiles/Drough_Impacts_Report_June2013_final.pdf.
Landau, Martin. 1969. Redundancy, rationality, and the problem of duplication and overlap. *Public Administration Review* 29: 346–358.
Larreguy, Horacio A. and Joana C.M. Monteiro. 2013. The role of media networks in compensating political biases: Evidence from radio networks in Brazil. Mimeo, Harvard.
Lebel, Louis, Elena Nikitina, Vladimir Kotov, and Jesse Manuta. 2006. Assessing institutionalized capacities and practices to reduce the risks of flood disaster. In: Birkmann, Joern (ed.), *Measuring Vulnerability to Natural Hazards: Towards Disaster Resilient Societies*. United Nations University Press, Tokyo, Japan, pp. 359–379.
Lemos, Maria C. 2007. Drought, governance, and adaptive capacity in northeast Brazil: A case study of Ceará. Occasional paper for UNDP 2007/2008. United Nations Development Program, Human Development Report Office, New York.
Lerner, Allan W. 1987. There is more than one way to be redundant. *Administration & Society*, 3: 334–337.
Lorenzo-Lacruz, J., Sergio M. Vicente-Serrano, Juan I. López-Moreno, Santiago Beguería, José María García-Ruiz, and José María Cuadrat. 2010. The impact of droughts and water management on various hydrological systems in the headwaters of the Tagus River (central Spain). *Journal of Hydrology*, 386, pp. 13–26.

Magalhães, Antônio R., H.C. Filho, F. L. Garagorry, J. G. Gasques, L. C. B. Molion, M. da S. A. Neto, C. A. Nobre, E. R. Porto, and O. E. Rebouças. 1988. The effects of climatic variations on agriculture in northeast Brazil. In: Parry, M. L., T. R. Carter, and N. T. Konijn (eds.), *The Impact of Climatic Variations on Agriculture*, volume 2, Assessments in Semi-Arid Regions. UNEP, IIASA, Kluwer Academic Publishers, Dordrecht, the Netherlands.

Magalhães, Antônio R. and Michael H. Glantz. 1992. *Socioeconomic Impacts of Climate Variations and Policy Responses in Brazil*. UNEP-SEPLAN Ceará-ESQUEL, Brasília, Brazil.

Magalhães, Antônio R. and Eduardo S. Martins. 2011. Drought and drought policy in Brazil. In: Sivakumar, Mannava V.K., Raymond P. Motha, Donald A. Wilhite, and John J. Qu (eds.), *Towards a Compendium on National Drought Policy. Proceedings of an Expert Meeting*, Washington, DC, July, pp. 57–67.

Magalhães, Antônio R. and Eduardo B. Neto. 1991. *Impactos Sociais e Econômicos de Variações Climáticas e Respostas Governamentais no Brasil*. Governo do Ceará-PNUMA. Fortaleza, Brazil.

Magalhães, Antônio R., Eduardo B. Neto, and Stahis Panagides. 1993. *Projeto Áridas: Políticas para o desenvolvimento sustentável no Nordeste*. Seplan-IICA-Esquel (Mimeo), Brasília, Brazil.

Marengo, José A. and Mauro Bernasconi. 2015. Regional differences in aridity/drought conditions over Northeast Brazil: Present state and future projections. *Climatic Change*, 129: 103–115.

Martins, Eduardo S.P.R. and Antônio R. Magalhães. 2015. *A seca de 2012–2015 no Nordeste e seus impactos, Revista Parcerias Estratégicas*, volume 20, No. 41, Brasília, Brazil, pp. 107–128.

Martins, Eduardo S.P.R., Cybelle F.C. Braga, Erwin De Nys, Francisco de A. de Souza Filho, and Marcos A. de Souza Freitas. 2013. *Impacts of Climate Change and Demand. Projections on the Process of Water Allocation in Two Basins in the Semi-Arid Northeast*, volume 8, Série Água Brasil. World Bank, Brasília, Brazil.

Mckee, Thomas B., Nolan J. Doesken, and John Kleist. 1993. The relationship of drought frequency and duration to times scale. In: *Proceedings of the 8th Conference on Applied Climatology*, volume 17, No. 22, American Meteorological Society, Boston, MA (Preprints), pp. 179–184.

Mckee, Thomas B., Nolan J. Doesken, and John Kleist. 1995. Drought monitoring with multiple time scales. In: *9th AMS Conference on Applied Climatology*. American Meteorological Society, Boston, MA, pp. 233–236.

Ministerio de Medio Ambiente. 2007. *Plan Especial de Alerta y Eventual Sequía em La Confederación Hidrográfica del Júcar*. 185p.

MNI—Ministério da Integração Nacional. 2015. www.integracao.gov.br. Brasília, Brazil.

Mueller, Valérie A. and Daniel E. Osgood. 2009. Long-term impacts of droughts on labour markets in developing countries: Evidence from Brazil. *Journal of Development Studies* 45(10): 1651–1662.

National Drought Mitigation Center (n.d.) Checklist of historical, current, and potential drought impacts. National Drought Mitigation Center. Available from: http://drought.unl.edu/portals/0/docs/checklist.pdf.

National Drought Policy Commission. 2000. *Final Report: Preparing for Drought in the 21st Century*. United States Department of Agriculture, Washington, DC.

Neves, Frederico de Castro. 1995. Curral de Bárbaros—Os Campos de Concentração no Ceará, 1915 e 1932. *Revista Brasileira de História*, 14(29): 93–122.

Organisation for Economic Cooperation and Development. 1993. *Core Set of Indicators for Environmental Performance Reviews—A Synthesis Report by the Group on the State of the Environment*, volume 83, Environmental Monographs. OECD, Paris, France, 39p.

Pahl-Wostl, Claudia. 2007. Transitions towards adaptive management of water facing climate and global change. *Water Resources Management*, 21: 49–62.

Painel Brasileiro de Mundanças Climáticas. 2013. *Impactos, Vulnerabilidades e Adaptação. Contribuição do Grupo de Trabalho 2 ao Primeiro Relatório de Avaliação Nacional do Painel Brasileiro de Mudanças Climáticas. Sumário Executivo do GT2*. PMBC, Rio de Janeiro, Brazil.

Pereira, L.S. and A.A. Paulo. 2004. Indicadores de Escassez de Água: Índices de Secas" in Conf. Conv. Workshop Indicadores y Tecnologias de Uso del Agua em las Tierras secas deIberoamérica (Proyeto CYTEDXVII-1, Junho 2004) CYTED, Mendoza, Argentina.

PES—Plan Especial de Alerta y Eventual Sequía em LA Confederación Hidrográfica del Júcar. 2007. Ministerio de Medio Ambiente. Mexico 185p.

PNUMA—Programa das Nações Unidas para o Meio Ambiente. 2007. *Projeto Geo Cidades: relatório ambiental urbano integrado - Rio de Janeiro*. Rio de Janeiro: PNUMA/MMA/IBAM/ISER/REDEH.

PROJETEC—BRLi. 2010. *Plano hidro ambiental da bacia hidrográfica do rio Capibaribe*, volume 01/3, Tomo I—Diagnóstico hidro ambiental. Projetos técnicos, Recife, Brazil, 389p.

Projeto Áridas. 1995. *A Strategy for Sustainable Development in Brazil's Northeast*. Ministry of Planning and Budget, Brasília, Brazil.

Revista Época. 1985. *Vidas Secas: São Paulo vive a maior crise hídrica em 50 anos. Culpa do governo ou de São Pedro?* Edição 886, Editora Globo, São Paulo, Brazil.

Rockstrom, Johan, Louise Karlberg, Suhas P. Wani, Jennie Barron, Nuhu Hatibu, Theib Oweis, Adriana Bruggeman, Jalali Farahani, and Zhu Qiang. 2010. Managing water in rainfed agriculture—The need for a paradigm shift. *Agricultural Water Management*, 97: 543–550.

Rosendo, Eliamin E.Q. 2014. Desenvolvimento de indicadores de vulnerabilidade à seca na região semiárida brasileira. Masters' Thesis, Programa de Pós-Graduação em Engenharia Urbana e Ambiental, Universidade Federal da Paraíba, João Pessoa-PB, Brazil.

Safi, Ahmad S., William J. Smith, and Zhnongwei Liu. 2012. Rural Nevada and climate change: Vulnerability, beliefs, and risk perception. *Risk Analysis*, 32(6). DOI: 10.1111/j.1539-6924.2012.01836.x.

Salvati, Luca, Marco Zitti, Tomaso Ceccarelli, and Luigi Perini. 2009. Developing a synthetic index of land vulnerability to drought and desertification. *Geographical Research*, 47-3: 280–291. DOI: 10.1111/j.1745-5871.2009.00590.x.

Sarmento, Francisco J. 2007. Mudancas climáticas: Impactos sociais, econômicos e ambientais no semi-árido brasileiro. In: *XVII Simposio Brasileiro de Recursos Hidricos*, São Paulo, Brazil.

Schaefer, Joseph T. 1990. The critical success index as an indicator of warning skill. *Weather and Forecasting*, 5: 570–575.

Scheper-Hughes, Nancy. 1992. *Death Without Weeping: The Violence of Everyday Life in Brazil*. University of California Press, Berkeley, CA.

Scott, W. Richard. 1985. Systems within systems. *American Behavioral Scientist*, 28: 601–618.

Sentelhas, Paulo C. 2010. Agricultural drought indices in current use in Brazil. In: Sivakumar, Mannava V.K., Raymond P. Motha, Donald A. Wilhite, and Deborah A. Wood (eds.), *Agricultural Drought Indices. Proceedings of an Expert Meeting*, Murcia, Spain, pp. 72–84.

Shukla, Shraddhanand and Andrew W. Wood. 2008. Use of a standardized runoff index for characterizing hydrologic drought. *Geophysical Research Letters*, 35: L02405, 7p. DOI: 10.1029/2007GL032487.

Smith, Herbert H. 2012. *Brazil: The Amazons and the Coast*. Forgotten Books, New York (originally published in 1879).

SRH—Secretaria de Recursos Hídricos. 2010. *Plano hidroambiental da bacia hidrográfica do rio Capibaribe*, volume 01/03, Tomo I—Diagnóstico hidroambiental. Projetos Técnicos, Recife, Brazil.

Svoboda, Mark. 2000. An introduction to the drought monitor. *Drought Network News*, 12(1): 15–20.

Svoboda, Mark, Doug LeComte, Mike Hayes, Richard Heim, Karin Gleason, Jim Angel, Brad Rippey et al. 2002. The drought monitor. *Bulletin of the American Meteorological Society*, 83(8): 1181–1190.

Taddei, Renzo and Ana L. Gamboggi. 2010. *Depois que a Chuva nao veio. Respostas Sociais às Secas no Nordeste, na Amazonia e no Sul do Brasil*. FUNCEME/CIFAS, Fortaleza, Brazil.

Thomas, D.S.K., Olga V. Wilhelmi, Taryn N. Finnessey, and Veva Deheza. 2013. A comprehensive framework for tourism and recreation drought vulnerability reduction. *Environmental Research Letters*, 8: 044004. DOI: 10.1088/1748-9326/8/4/044004.

Tsegai, Daniel and Reza Ardakanian. 2014. *Proceedings of the Regional Workshop on Capacity Development to Support National Drought Management Policies for Latin America and the Caribbean Countries*. UN-Water Decade Programme on Capacity Development (UNW-DPC).

Turner II, Benjamin L., Roger E. Kasperson, Pamela A. Matson, James J. McCarthy, Robert W. Corell, Lindsey Christensen, Noelle Eckley et al. 2003. A framework for vulnerability analysis in sustainability science. *Proceedings of the National Academy of Sciences of the United States of America*, 100(14): 8074–8079.

United Nations International Strategy for Disaster Risk Reduction. 2004. *Living with Risk: A Global Review of Disaster Reduction Initiatives*, volume 1, International Strategy for Disaster Reduction. United Nations Publications, Geneva, Switzerland.

United States Geological Survey. 2005. *Shuttle Radar Topography Mission (SRTM)*. Projeto SRTM/NASA. USGS. Available from: http://edc.usgs.gov/products/elevation/srtm. Last Accessed on November 15, 2015.

Varvasovszky, Zsuzsa and Ruairí Brugha. 2000. How to do (or not to do)... A stakeholder analysis. *Health Policy and Planning*, 15(3): 338–345.

Vergara, Walter and Sebastian M. Scholtz. 2011. *Assessment of the Risk of Amazon Dieback*. World Bank, Washington, DC.

Verner, Dorte. 2010. *Reducing Poverty, Protecting Livelihoods, and Building Assets in a Changing Climate*. World Bank, Washington, DC.

Verner, Dorte and Edinaldo Tebaldi. 2015. *Drought, Migration and Social Policies in Drylands of the Middle East and Latin America*. Mimeo, New York.

Vicente-Serrano, Sergio M., Santiago Beguería, and Juan I. López-Moreno. 2010. A multiscalar drought index sensitive to global warming: The standardized precipitation evapotranspiration index—SPEI. *Journal of Climate*, 23: 1696–1718.

Vicente-Serrano, Sergio M., Santiago Beguería, Jorge Lorenzo-Lacruz, Jesus J. Camarero, Juan I. López-Moreno, Cesar Azorin-Molina, Jesús Revuelto, Enrique Morán-Tejeda, and Arturo Sánchez-Lorenzo. 2012. Performance of drought indices for ecological, agricultural and hydrological applications. *Earth Interactions*, 16: 1–27.

Voltolini, Ricardo. 2004. *Terceiro setor: Planejamento e gestão*. Editora Senac, São Paulo, Brazil.

Wilhite, Donald A. ed. 1993. Planning for drought: A methodology. In: *Drought Assessment, Management, and Planning: Theory and Case Studies*. Kluwer Academic Press, Boston, MA, pp. 87–108.

Wilhite, Donald A. 1996. A methodology for drought preparedness. *Natural Hazards*, 13: 229–252.

Wilhite, Donald A. ed. 2000. Drought as a natural hazard: Concepts and definitions. In: *Drought: A Global Assessment*, volume 1. Routledge, London, UK, pp. 3–18.

Wilhite, Donald A. and Margie Buchanan-Smith. 2005. Drought as hazard: Understanding the natural and social context. In: Wilhite, Donald A (ed.), *Drought and Water Crises: Science, Technology, and Management Issues*. CRC Press, Boca Raton, FL, pp. 3–29.

Wilhite, Donald A. and Michael H. Glantz. 1985. Understanding the drought phenomenon: The role of definitions. *Water International*, 10(3): 111–120.

Wilhite, Donald A., Michael J. Hayes, and Cody L. Knutson. 2005. Drought preparedness planning: Building institutional capacity. In: Wilhite, Donald A (ed.), *Drought and Water Crises: Science, Technology, and Management Issues*. CRC Press, Boca Raton, FL, pp. 93–135.

Wilhite, Donald A., Mannava V.K. Sivakumar, and Roger Pulwarty. 2014a. Managing drought risk in a changing climate: The role of national drought policy. *Weather and Climate Extremes*. Available from: http://dx.doi.org/10.1016/j.wace.2014.01.002.

Wilhite, Donald A., World Meteorological Organization (WMO), and Global Water Partnership (GWP). 2014b. *National Drought Management Policy Guidelines: A Template for Action*. Integrated Drought Management Programme (IDMP) Tools and Guidelines Series 1. WMO, Geneva, Switzerland, and GWP, Stockholm, Sweden.

Wisner, Ben. 2006. Self-assessment of coping capacity: Participatory, proactive and qualitative engagement of communities in their own risk management. In: Birkmann, Joern (ed.), *Measuring Vulnerability to Natural Hazards: Towards Disaster Resilient Societies*. United Nations University Press, Tokyo, Japan.

World Bank. 2003. *Sustainable Development in a Dynamic World: Transforming Institutions, Growth, and Quality of Life*, World Development Report 2003. World Bank, Washington, DC.

World Bank. 2005. *Drought in the Amazon – Scientific and Social Aspects*. A World Bank Seminar, Brasília, Brazil.

World Bank. 2010. *Development and Climate Change*, World Development Report 2010. World Bank, Washington, DC.

World Bank. 2012. *The Sendai Report: Managing Disaster Risks for a Resilient Future*. World Bank, Washington, DC. Available from: http://www.gfdrr.org/sites/gfdrr.org/files/Sen dai_Report_051012.pdf.

World Bank. 2014. *Turn Down the Heat: Confronting the New Climate Normal*. World Bank, Washington, DC.

World Bank. 2015a. *Protocolo de Preparação para a seca da Bacia Hidrográfica do Rio Piranhas-Açu.* Relatório Final (in press).

World Bank. 2015b. *Relatório Geral do Monitor*, Brasília, Brazil.

World Bank. 2015c. *Shock Waves: Managing the Impacts of Climate Change on Poverty.* World Bank, Washington, DC.

World Meteorological Organization (WMO). 2013. High level meeting on national drought policy (HMNDP), Geneva. Available from: http://www.droughtmanagement.info/literature/WMO_HMNDP_final_declaration_2013_en.pdf.

Ximenes, Luciano J.F., P. Lopes, and E. Mendonca. 2013. *Efeito das secas sobre indicadores agropecuarios do Nordeste.* Escritorio Tecnico de Estudos Economicos do Nordeste, Banco do Nordeste.

Index

Note: Page numbers followed by f and t refer to figures and tables, respectively.

A

Açudes, 10–11
Adaptive process, 51, 61
Agentes de Saúde (community health agents), 104–105
Agreste region, 2, 4–5, 61, 193
Agricultural production
  average local impact on, 135
  climate change impacts on, 99
  coefficients, 135, 135t
  data on, 123
  gross value of, 129, 130t, 131t, 132t
  logarithms of, 125
  losses in, 136
  World Bank study, 93
Amazon
  dieback, 7
  drought in, 7
  roads, 12
  SIPAM, 24f
ANA. *See* National Water Agency (ANA)
Andreu, Joaquín, 85–86
Animal-related production, 138, 138t
Aridas Project, 14
Aridity, 1, 82
Authorship meeting, 158–159

B

Banabuiú River Basin, 69
Basin DPPs, 174–190
  actions, 189–190, 191t–192t
  context, 174, 178–179
  indicators and drought monitor for, 179–188
    DM-PARB, 186–188
    NDM, 180
    SI, 185–186
    SPI, 182–183
    SRI, 183, 185
  vulnerability analysis, 188–189
Bolsa Estiagem program, 25
Bolsa Família program, 16, 25, 106–107
Brazil. *See also* Northeast Brazil
  drought in. *See* Drought(s)
  hydrological crisis, 19–22
  life in, 1–18
  map of, 3f
  river basin, 55
Brazilian Government mission, 26
Brazilian Institute of Geography and Statistics (IBGE), 123

C

Caatinga vegetation, 3, 5, 11
Cantareira system, 6, 20f
Casa Civil, 29, 139
Ceará, 8, 102, 105
  Fortaleza metropolitan system in, 59, 61
  intra- and inter-year distribution of rainfall, 37–38, 38f
  municipality of Piquet Carneiro, 68–69
  water storage capacity of, 49, 50f
Ceará Foundation for Meteorology and Water Resources (FUNCEME), 43, 46
Ceará Water Resource Management Company (COGERH), 202
Central Institution (CI), 155–158, 161
Chronogram, 158
Climate change
  adaptation plan, 94
  adding, 92–93
  and droughts in Northeast Brazil, 17
  hydrological effects of, 93
  impacts on overall economy, 99
  northeast municipalities, 98–99
  resilience, 91–95

Climate-driven migration (*sertão*), 106–107, 117f
COMDEC (Municipal Civil Defense Coordination Unit), 71–74, 77
Company for the Development of the São Francisco and Parnaíba Valleys (CODEVASF), 11–12
CONAGUA (Mexican Water Authority), 84
Contingency analysis method, 172–173
Convening power, 29–31
"Creeping" phenomenon, 92
Crisis
  case of São Paulo, 20
  management, 22–29
  overview, 19–22

D

Data sources, 123–124
  on agricultural production, 123
  blend indicators, droughts, 124
Decision-making process, 73, 78
Decision support system (DSS), 50, 83, 207
Degree of drought severity, 173t
Desertification, 1–2
Detrimental effect, 97
Disaster risk management (DRM), 30
  cycle, 28–29, 28f
  in five pillars, 32
Dissemination plan, 70
Distress migration, climate-induced, 106–107
DM-PARB. *See* Drought Monitor for the Piranhas-Açu River Basin (DM-PARB)
DNOCS. *See* National Department for Works Against Droughts (DNOCS)
Drought(s), 49
  definition, 5
  econometric estimates on economic impacts of, 135–136, 135t, 138–139
  future, 94
  indicators. *See* Indicators, drought
  indices, 168–169
    evaluation, 172–173
    shortcomings of, 172
  management, 174
    institutions in, 24f
    integrating, 52
    operational plans, 52
    for Piranhas-Açu River Basin, 191t–192t
    thematic axis, 58t
  monitoring system, 60
  opportunity for change, 22–23
  periods. *See* Periods, drought
  planning
    adding climate change to mix, 92–93
    approaching, 94–95
    conceptual framework, 51–53
    as instrument, 51
    monitoring, 52–53
    on multiple spatial, temporal, and sectoral scales, 53–55
    opportunity, 91–92
    risk and impact assessment, 53
    scales of, 54f
    state drought plan, 54
  policies and DPPs, 94
  policy response to current, 139–141, 141t
  refugees, 8
  risk management, 72–73
  risks, 50, 52
  on set of outcomes, 125
  severity, 152f, 164f, 165f
  system failures, 52
  traditional measures, 25
  variables, 122
  ways to assist people during, 14–15
Drought Monitor. *See also* Monitor Map
  actors involved, responsibilities, and users, 155–157
    authors, 156
    CI, 155–156
    institutions providing data, 155
    results, 156
    technical group for data quality and IT, 156–157
    validators, 156
  challenges and use of, 47–48
  characteristics, 41–42
  concepts in designing, 153–155
  DPPs and, 45–47

drought categories and impacts using, 42t
drought indicators. *See* Indicators, drought
final map, 162f
flowchart of activities to update, 157–165
data preparation, 157–158
design, 158–159
map publication, 161–165
validation, 159–161
impact lines, 154
information technology and data utilized, 143–144
institutional cooperation, support of, 43, 45, 47
layer with reservoir conditions, 152–153
layout of Monitor site, 163f
model, 40
Northeast Brazil, experimental map, 44f
operational arrangements of, 155–165
operational implementation of, 41
operational scheme for, 157f
overview, 144–145
solution based on, 39–45
traditional and conventional monitoring difference, 40
traditional approach, limitation, 37–39
Ceará, intra- and inter-year distribution of rainfall, 37–38, 38f
drought management, 38–39
Drought Monitor for the Piranhas-Açu River Basin (DM-PARB), 186–188
Drought Observatory, 24–25
Drought preparedness plans (DPPs), 50, 168, 201t
basin, 174–190
first steps of change, 33–34
for hydrosystems, 53f, 55, 59–60
mitigation and response strategies, 60
operational planning for, 59–60
surface water supply sources, 59
water allocation measures, 59
for irrigation schemes, 55
municipal agricultural response plan and, 70–76, 71f
background, 67–69
development, 70
diagnosis, 70
dissemination plan, 70
drought events, 74–75
drought response and mitigation activities, 73–74
drought risk management, 72
Piquet Carneiro. *See* Piquet Carneiro
planning, 70
reasons, 70
timeline for preparation of, 71f
for rain-fed agriculture, 55
for river basin, 53f, 54–58, 56f
assessment of impacts and vulnerability, 56
categories, 57
drought scenarios, 57t
indicators, 57
mitigation and response actions, 56
monitoring, 56
water resource management institutions, 56
three pillars, 32–33
mitigation and response planning, 33
monitoring and forecasting/early warning, 32–33
vulnerability/resilience and impact assessment, 33
urban, 202–208
for urban water utilities, 53f, 55, 61–64
elements, 63
as evolutionary process, 61, 62f
legal framework, 61
measures, 63–64
methodological steps for elaboration of, 62–63, 62f
for water supply to dispersed rural populations, 55
Drought Preparedness Protocol, 178–179, 180f, 188, 190, 190t
Drought Prevention and Mitigation Programs (PMPMS), 87

Drought-related migration, 107
Dry spell indicators, 148–150
DSS. *See* Decision support system (DSS)

E

Econometric estimates on economic impacts of drought, 135–136, 135t, 138–139
  animal-related production, 138, 138t
  livestock types, 136, 137t
  losses in agricultural production, 136
Economic implications, 102–104
Economic indicators, trends in main, 129
Elite capture problem, 10
El Niño phenomenon, 5, 16
EMATERCE monitoring system, 77
Empirical methodology, 124–125
Empirical results, 125–139
  econometric estimates on drought, 135–136, 135t, 138–139
    animal-related production, 138, 138t
    livestock types, 136, 137t
    losses in agricultural production, 136
  trends
    in drought indicators, 125, 126t, 127t, 128
    in main economic indicators, 129
    in volume of water reservoirs relative to capacity, 128–129, 128t
Environmental impact measures, 64
Evolutionary planning process, 51, 61
Ex ante risk management, 73–74

F

False alarm ratio (FAR), drought, 172–173
Fixed effect econometric methods, 122
Flagelados, 8
FONDEN (Natural Disaster Fund), 86
Forest zone, 2
Fortaleza metropolitan region (FMR), 202, 204
  vulnerability assessment methodology, 202
  water treatment plant, 204, 205f
Fortaleza metropolitan system in Ceará, 59, 61
Frentes de trabalho, 14, 16
FUNCEME (Ceará Foundation for Meteorology and Water Resources), 43, 46
Furtado, Celso, 12

G

Gamma distribution, 170
Gandu, Adilson, 153
Garantia-Safra (harvest guarantee), 25
Geographic Information System (GIS), 155
Getúlio Vargas Camp, 8
Gonzaga, Luiz, 9
Gross domestic product (GDP), 129, 134t
  agricultural value added and, 138–139, 139t
  of Piquet Carneiro, 69
Growth Acceleration Program (PAC), 140
Guide for Elaboration of Emergency Drought Plans in Urban Water Supply Systems, 63

H

Health implications, 104–105
High-Level Meeting on National Drought Policy (HMNDP), 29–30, 33
Hydrographic Confederation of the Jucar (CHJ), 83
Hydroillogical cycle, 6, 51
Hydrological crisis, drought, 19–22
  2010–2015, 20–22
  characteristics, government responses, 19
  in São Paulo, 20
  southeast region, 20–21
Hydrological drought indicator, 170–171
Hydrological Planning Unit (HPU), 180
  Médio Piranhas Paraibano, SRI-6, 185f
  Piancó, state index, 185–186, 186f
  Seridó, 183f, 188–189

Hydrosystems, DPPs, 52, 53f, 55, 59–60
mitigation and response strategies, 60
operational planning for, 59–60
analysis of legal and institutional context, 59
decision making, 60
drought category, 60
drought monitoring system, 60
risk tolerance level and drought phases, 60
simulation and optimization models, 60
surface water supply sources, 59
water allocation measures, 59

I

IIAMA (Institute for Water and Environmental Engineering), 82
IMTA (Mexican Institute of Water Technology), 84
Incomplete gamma function, 170
Indicators, drought, 122, 145–153, 156, 158
blend, 124
layer with reservoir conditions, 152–153
representation of products, 151f
SPEI, 148
SPI, 146–147
SRI and SDSI, 148–150
support products, 150–151
trends in, 125, 126t, 127t, 128
and triggers, 168–174
applications, 174, 175t–179t
based on storage level, 171–172
hydrological, 170–171
meteorological, 169–170
overview, 168–169
performance evaluation, 172–174
vulnerability, 188f
Infant mortality rate (IMR), 104
Institutional measures, 64
Integrated Committee to Combat Drought, 25
Irrigation schemes, DPPs, 55
IS. *See* Synthetic index (IS)

J

Jucar River Basin, 83, 85
Jucazinho water system, 174, 193
in Pernambuco, 59, 61, 63, 168, 174
reservoir simulation, 193–201
histogram of severity of operations, 198f–199f
permanence curve, 200f
standard operation, 194, 198f
target operation, 194, 195f–196f, 196t, 197f, 199t
values of
drought indices, 175t–177t
FAR and POD, 178t, 179t

L

Labor migration, climate-induced, 106–107
Legal and normative measures, 64
Literatura de cordel, 9
Livelihood implications, 102–104
Livestock, types, 136, 137t

M

Meteorological drought indicator, 169–170
Methodology for the Design of Drought Monitor Maps for the Northeast, 153
Mexican Institute of Water Technology (IMTA), 84
Mexican National Drought Policy, 84
Mexican Water Authority (CONAGUA), 84
Mexico, paradigm shift, 83–84
Migration, 106–117
climate-driven, 106–107, 117f
drought-related, 107
seasonal, 107
Ministry of National Integration (MNI), 11, 23–26, 30, 40, 101
Mitigation measures, drought, 64, 201
monitoring, 202
organizational, 201
and response planning, 33
Monitor Map, 145, 154–155, 159f–160f

Muddy water, 101
Municipal agricultural DPPs and response plan, 70–76, 71f
  background, 67–69
  development, 70
  diagnosis, 70
  dissemination plan, 70
  drought events, 74–75
  drought response and mitigation activities, 73–74
  drought risk management, 72
  Piquet Carneiro. *See* Piquet Carneiro
  planning, 70
  reasons, 70
  timeline for preparation of, 71f
Municipal Agricultural Production (PAM), 123
Municipal Alert Index, 68–69
Municipal Civil Defense Coordination Unit (COMDEC), 71–74, 77
Municipal Development Index, 68

N

National Civil Defense Secretariat, 23, 25
National Department for Works Against Droughts (DNOCS), 11
National drought management policies, 29, 34
National Drought Mitigation Center (NDMC), 75, 84–85
National Drought Policy, 47, 145
  development in Mexico, 89–90
  elaboration of, 26
  need for, 40
  overarching principles, 33
National Program against Drought (PRONACOSE), 86
National Water Agency (ANA), 30, 45–46, 155
National Water Law, 23
National Water Resource Management System, 59
National Water Resources Policy, 59
National Water Security Plan, 94
Natural Disaster Fund (FONDEN), 86
Natural/physical drought, 154, 156
Nocturnal interruption, 207
North American Drought Monitor (NADM), 84
Northeast Brazil, 2–4, 123
  agricultural production, 129, 130t, 131t, 132t, 133t
  data sources, 123–124
    on agricultural production, 123
    blend indicators, droughts, 124
  degrees of drought severity, 164f
  diseases in, 105
  drought in, 4–6
    1877–1879 period, 8, 10
    2010–2015, 16–17, 22, 24
    causes of periodic, 5
    Ceará, 8
    climate change and, 17
    emergency responses to, 14–15
    government's response for, 10–14
    houses during, 105, 116f
    impacts of, 7–10
    policy response to current, 139–141, 141t
    proactive response, 17–18
    reactive policies, 10–14
    records of, 5
    São Paulo, 6
  Drought Monitor, experimental map, 44f
  eco-regions of, 4f
  empirical methodology, 124–125
  empirical results, 125–139
    economic impacts of drought, 135–136, 135t, 138–139
    trends, 125–129
  fertility rate, 105
  GDP growth rates, 129, 134t
  government initiatives, 67–68
  MNI and partners, 30, 31f
  pertinent literature, 120–123
    descriptive case studies, 120–121
    econometric studies, 120–122
  rain-fed agriculture, 67
  states of, 3f
  water crisis in, 19, 21–22
  water infrastructure investment, 38–39

Northeast Drought Monitor (NDM), 34, 40, 47. *See also* Drought Monitor
blend indicators, 124, 142
constitution, 43
development process, 88
for Piranhas-Açu River Basin, 180, 187
Northeast municipalities, climate change, 98–99

O

Operational drought, 52
Operational planning for hydrosystem
analysis of legal and institutional context, 59
decision making, 60
drought category, 60
drought monitoring system, 60
risk tolerance level and drought phases, 60
simulation and optimization models, 60

P

PAC (Growth Acceleration Program), 140
Paradigm shift, 28–29
supporting Brazilian, 85–88
Periods, drought
1877–1879, 8, 10, 18
2010–2015, 16–17, 20–22, 24, 67
multiyear, 19, 21
problems in, 21
carros pipas (tank trucks), 101
in Ceará (2012 to present), 70
drought-stricken communities, 104
impacts on agricultural sector, 75
livestock, 104
trees, 104
Permanence curve, water system operation, 196, 199, 200f
Permanent Drought Committee (PDC), 83
Pernambuco state, 152, 156
drought severity in, 152, 152f
Jucazinho system in, 59, 61, 63, 168, 174
volume of water reservoirs, 128
Pertinent literature, 120–123
descriptive case studies, 120–121
econometric studies, 120–122
Peti (Child Labor Eradication Program), 106
Pillars of drought preparedness, 32–33
Pilot program, 45–46
Piquet Carneiro
characterization of, 69–70
economy of, 69
effectiveness and sustainability of plan, 76–77
identification of key stakeholders, 76
information systems and continuous monitoring, 78–79
inter-institutional arrangements and involvement of private sector, 79–80
leadership support, 77
managing expectations and funding, 77
model for replication of new plans, 78
municipal agricultural preparedness and response plan, 70–76, 71f
municipality of, 68
population of, 69
problem tree analysis, 74f
promoting socioeconomic development, 73
technical guidance and capacity building, 78
Piranhas-Açu (PA) river basin, 55, 58, 178, 180f, 186–188
DPPs, 56f, 182f
drought management
actions for, 191t–192t
thematic axis, 58t
drought monitor for, 187t
Drought Preparedness Protocol, 190t
state index, 186t
vulnerability, 188, 189t
water basin of, 181f
PMPMS (Drought Prevention and Mitigation Programs), 87
Policy response to current drought, 139–141, 141t
Political–institutional momentum, 50

Pre-Amazônia, 3
Preparedness measures, 64
Pressure-state-impact-response (PSIR) methodology, 202–204, 205f
Proactive drought management, 91
Probability distribution (gamma distribution), 146
Probability of detection (POD), drought, 172–173
Programa de Leite (milk program), 105
PRONACOSE (National Program against Drought), 86
Publication process, 164

R

Rain-fed agriculture, 17, 124, 136
  DPPs, 55
  drought preparedness and response plan, 68
  for small-scale farmers, 67
Reactive policies for drought, 10–14
  elite capture problem, 10
  engineers commission, 10–11
  institution, 11
  integrated water resource management, 13
  New Republic government actions, 13
  scientific commission, 10
  world banks in, 12
Relative drought, 153
Reservoirs
  failure, severity and vulnerability, 194
  layer with conditions, 152–153
  relative to capacity, trends, 128–129, 128t
  simulation, Jucazinho water system, 193–201
    histogram of severity of operations, 198f–199f
    permanence curve, 200f
    standard operation, 194, 198f
    target operation, 194, 195f–196f, 196t, 197f, 199t
Resilience
  reservoir, 194, 203
  time of Jucazinho water system, 199f
Retirantes, 8
Rio Grande do Norte Agriculture Research Center (EMPARN), 157
Risk management, 29, 51–52
Risk monitoring measures, 64
River basin. *See also specific river basins*
  defined, 55
  DPPs, 53f, 54–58, 56f
    assessment of impacts and vulnerability, 56
    categories, 57
    drought scenarios, 57t
    indicators, 57
    mitigation and response actions, 56
    monitoring, 56
    water resource management institutions, 56
Rural households, demand for water, 101
  private properties, cisterns, 111f
  traveling to collect water, 101, 109f, 110f
  for washing clothes, 110f
Rural pensions program, 106–107

S

São Francisco Inter Basin Water Transfer Project, 94
São Francisco River, 10–11
São Jose III Project, 76, 80
São Paulo
  Cantareira system in, 19
  crisis in (2014–2015), 20
  drought in, 6
Seasonal migration, 107
Semiarid region, 2–4
Sertão, drought, 2, 5–6, 9, 61. *See also* Ceará; Northeast Brazil
  beans during, 113f, 114f
  cisterns in, 111f
  climate-driven. *See* Climate-driven migration (*sertão*)
  community health agents in rural settlement, 116f
  deplete assets and harm health, 108f
  distress migration, 107
  farmers in, 112f
  infant mortality, 104
  life in, 104, 115f

multiyear, 108f
municipal seat's water storage, 109f
people at truck stop, 117f
rural households. *See* Rural households, demand for water
stricken field, 115f
Smith, Herbert, 8
Social impact measures, 64
Social protection, 106
Socioeconomic development measures, 22
Southeast region, drought in, 19, 20f
Spain, 82–83
DSSs, 83
hydrological year 2004/2005, 83
IIAMA, 82
Jucar River Basin, 83
SDPs, 83
SODMIs, 83
UPV, 82
Special drought plans (SDPs), 83
Standardized dry spells indicator (SDSI), 149
Standardized operative drought monitoring indicators (SODMI), 83
Standardized precipitation-evapotranspiration index (SPEI), 147t, 148
calculation, 170
classifications, categories, and stages, 171t
disadvantages, 148
examples of drought intensity levels, 147f
uses of, 148
Standardized precipitation index (SPI), 146–147, 147t, 169, 182–183
advantages/disadvantages, 147
calculation, 169–170
classifications, categories, and stages, 171t, 182t
Drought Preparedness Protocol, 190t
examples of drought intensity levels, 147f
for Peixe and Seridó HPU, 182–183, 183f, 184t
probability distribution (gamma distribution), 146
Standardized runoff indicator (SRI), 149, 150f, 170, 183, 185
classifications, categories, and stages, 171t, 172t
State index (SI), 169, 171, 185–186
categorization of, 172t
Coremas-Mãe d'água in Piancó HPU, 185–186, 186f
Steering Committee, 43
Strategic Line of Intervention No. 1 (LE1), 71
Strategic Line of Intervention No. 2 (LE2), 71–72
Strategic Line of Intervention No. 3 (LE3), 72
Strategic Line of Intervention No. 4 (LE4), 72
Superintendency for Development of the Northeast (SUDENE), 11–13
Svoboda, Mark, 88
Synthetic index (IS), 171, 172t
System operation and management measures, 64

T

Technical Assistance and Rural Extension Enterprise of Ceará (EMATERCE), 73, 77–78
Technical collaboration program, 30, 34, 68
Technical Group (Monitor), 155–157
Tolerable risk levels (TRLs), 52
Trends
in drought indicators, 125, 126t, 127t, 128
in main economic indicators, 129
in volume of water reservoirs relative to capacity, 128–129, 128t
"Turn Down the Heat Report," 92

U

United Nations Office for Disaster Risk Reduction (UNISDR 2004), 79
The United States, paradigm shift, 84–85
Universidad Politécnica de Valencia (UPV), 82

Urban DPPs, 202–208
action plan, 206–207
indications for drought stage, 208t
management aspect, 207
normative framework, 206–207
to reduce water drought effect, 206
DSS, 207
monitoring and early warning, 203, 207
overview, 202–203
vulnerability analysis, 203–206
water utilities, 53f, 55, 61–64
elements, 63
as evolutionary process, 61, 62f
legal framework, 61
measures, 63–64
methodological steps for elaboration of, 62–63, 62f
US Drought Monitor (USDM) process, 85

V

Vidas Secas (*Dry Lives*), 9
Viva Criança (Long Live Children) program, 105
Vulnerability analysis
basin DPP, 188–189
urban DPP, 203–206

W

Water availability infrastructure measures, 64
Water management strategies, 94
Water reservoirs relative to capacity, trends in, 128–129, 128t
Water Resources Engineering Research Group (WRERG), 83, 86
Water scarcity, 100–101
Water storage capacity of Ceará, 49, 50f
Water supply to dispersed rural populations, DPPs, 55
Water system, drought plan, 190–202
actions, 201–202
operation, 193–201
overview, 190, 193
Water treatment plant (WTP), 204, 206
Wilhite, Donald, 6, 84
Work fronts, 13–16, 18, 25
Working group, MNI, 25–26
World Bank, 7, 30, 81–82, 86–87
defines DRM pillars, 32
drought policy and management, 81
on drought preparedness and climate resilience, 26, 30
learning exchange, 82
NDMC assistance, 87
rural development interventions, 12
system of conditional cash transfers, 16

Z

Zona da mata, 2–3, 61

For Product Safety Concerns and Information please contact our EU representative GPSR@taylorandfrancis.com Taylor & Francis Verlag GmbH, Kaufingerstraße 24, 80331 München, Germany

Batch number: 10726532

Printed by Printforce, the Netherlands